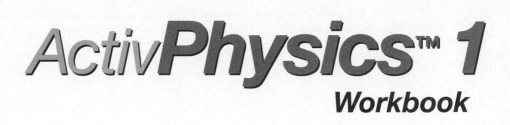

*Activ*Physics™ 1
Workbook

Alan Van Heuvelen
The Ohio State University

Addison Wesley Interactive

Director: Chip Price
Producer: Denise Descoteaux
Project Manager: Arati M. Nagaraj
Production Manager: Lee Stayton
Manufacturing Coordinator: Beverly Brissette
Workbook Design & Production: Dusan Koljensic
Production Assistance: Gabe Weiss
Copy Editing: Barbara Mindel

Library of Congress Cataloging-in-Publication Data

Van Heuvelen, Alan
    ActivPhysics 1 [computer file] / Alan Van Heuvelen. — Version 1.0.
    1 computer laser optical disc : 4 3/4 in. + 1 user's guide + 1 workbook
    Computer data and program.
    System requirements for Windows: 486/66Mhz or better; 8MB for Windows95 or 16MB for
Windows NT 4.0; Windows 95 or NT 4.0; 640X480 resolution, 256-color monitor; double-speed
CD-ROM drive; eight-bit, QuickTime-compatible sound card, speakers; mouse.
    System requirements for Macintosh: 68040/25 MHz Macintosh or Power Mac; 16MB RAM ;
System 7.5 or higher; 640x480 resolution, 256-color monitor; double-speed CD-ROM drive.
    Title from disc label.
    Summary: Multimedia introduction to physics covering mechanics, thermodynamics and waves.
Includes 122 guided activities, 115 simulations, animations, Java tools, video clips, and audio tracks.

    ISBN 0-201-84620-9
    ISBN 0-201-69480-8 (workbook)

    1. Physics--Software.   I. Title   II. Title: ActivPhysics one.   III. Title: Activ physics 1.

QC21.2"1996 01510""MRC"
530--DC12a   96-47216
                                          CIP

ISBN 0-201-69480-8

1 2 3 4 5 6 7 8 9 10 CRW 00999897

10. WAVES

1

DESCRIBING MOTION

A motion diagram uses a series of dots and arrows to represent the changing position, velocity, and acceleration of an object.

Advantages of Motion Diagrams:

* They help you develop mental images and intuition about the meaning of the kinematics quantities used to describe motion.

* They help you understand the signs of these quantities, especially when the quantities have negative signs.

* They are useful for checking the values of kinematics quantities when you are solving problems.

Examples of Motion Diagrams for Constant Acceleration (positive direction toward the right):

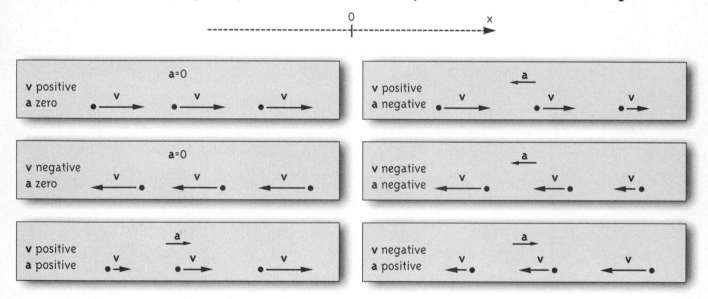

Rules for Constructing Motion Diagrams:

* The position dots indicate the location of the object at equal time intervals.

* The separation of adjacent position dots indicates roughly the speed of the object.

* $a = \Delta v / \Delta t$, and the direction of the acceleration arrow represents the change in the velocity Δv from one position to the next.

* The sign of the velocity or acceleration depends on the direction of the arrow relative to the coordinate axis (the positive direction is toward the right in these examples).

Question 1 — Meaning of x_0 : Set the initial velocity and acceleration sliders to zero and try different initial position settings. What is the meaning of x_0?

Question 2 — Meaning of v_0: Set the initial position and acceleration sliders to zero and try different initial velocity settings. What is the meaning of v_0 and v?

Question 3 — Meaning of a: Set the initial position to –48 m and the initial velocity to zero. Try different positive acceleration settings. How does the change in velocity each second relate to the acceleration?

Question 4 — Acceleration and Time: Set the initial position to –48 m and the initial velocity to + 12 m/s. Try different negative accelerations starting at – 1.0 m/s². Predict the time interval needed to stop the car.

Question 5 — Meaning of Negative a if v Is Negative: Set the initial position to +48 m and the initial velocity to zero. Try different negative accelerations. What does negative a imply about the motion if the object has a negative velocity? Start with a = –1.0 m/s².

Question 6 — Meaning of Positive a if v is Negative : Set the initial position to +48 m and the initial velocity to –12.0 m/s. Try different positive accelerations. Predict the time interval needed for the car to stop.

For each problem in Question 7, first run the simulation. Then adjust the initial position, the initial velocity, and the acceleration sliders so that the car has the same motion as the white dot. You should get the signs of the quantities correct on the first try but may need to experiment to get the exact values correct. After matching the motion, draw a motion diagram as a reminder of the motion that occurred.

Problem 1:

Initial slider-setting predictions:

$x_0 =$

$v_0 =$

$a =$

Slider settings that matched motion:

$x_0 =$

$v_0 =$

$a =$

Motion diagram that describes the motion:

Problem 2:

Initial slider-setting predictions:

$x_0 =$

$v_0 =$

$a =$

Slider settings that matched motion:

$x_0 =$

$v_0 =$

$a =$

Motion diagram that describes the motion:

Problem 3:

Initial slider-setting predictions:

$x_0 =$

$v_0 =$

$a =$

Slider settings that matched motion:

$x_0 =$

$v_0 =$

$a =$

Motion diagram that describes the motion:

Problem 4:

Initial slider-setting predictions:

$x_0 =$

$v_0 =$

$a =$

Slider settings that matched motion:

$x_0 =$

$v_0 =$

$a =$

Motion diagram that describes the motion:

The motion of a car is represented by motion diagrams and graphs. You can choose the motion by adjusting sliders for

- the initial position x_o
- the initial velocity v_o
- the acceleration

Answer the following questions and check your work by adjusting the sliders and running the simulation. The graphs that you are asked to draw are qualitative — don't worry about the detailed numbers.

Question 1— Initial Position: Describe in words the meaning of the quantity "initial position x_o."
Draw an x-vs-t graph for $x_o = -8$ m, $v_o = 0$, and $a = 0$.

Question 2 — Velocity and Changing Value of x: Draw x-vs-t and v-vs-t graphs for $x_o = 0$, $v_o = -4$ m/s and $a = 0$.
Predict the position x readings at $t = 0$ s, 1 s, 2 s, and 3 s.

Velocity and Position-Versus-Time Graph: Draw the following x-vs-t and v-vs-t graphs.

(a) $x_o = 0$, $v_o = +6$ m/s, $a = 0$

(b) $x_o = 0$ m, $v_o = +4$ m/s, $a = 0$

(c) $x_o = 0$, $v_o = +2$ m/s, $a = 0$

(d) $x_o = 0$ m, $v_o = -6$ m/s, $a = 0$

How is the slope of the x-versus-t graph related to the velocity?

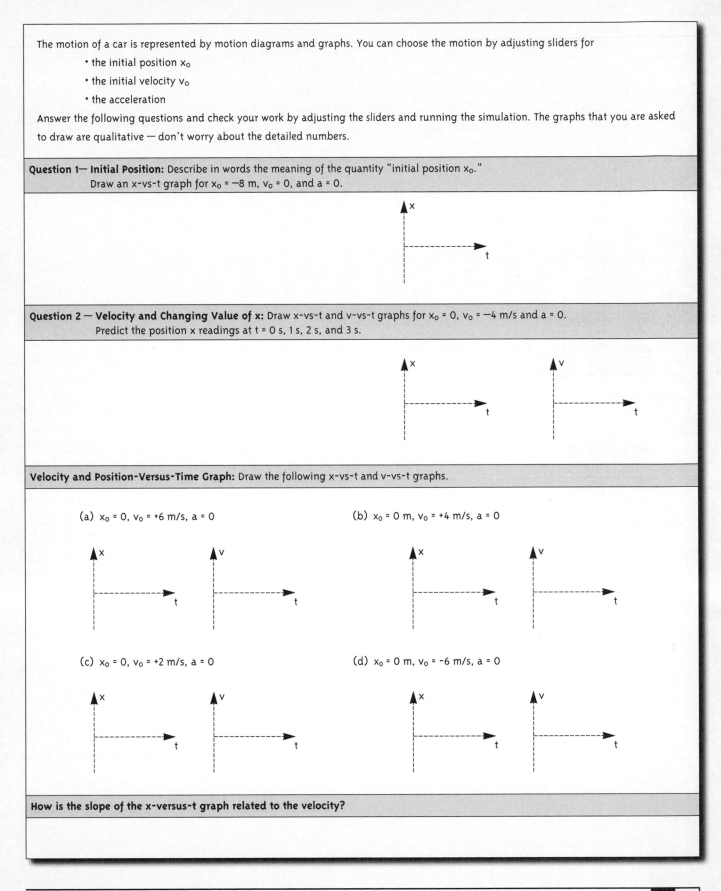

Questions 3-4 — Velocity and Acceleration-Versus-Time Graphs: Draw the following graphs.

(a) $x_0 = -48$ m, $v_0 = 0$ m/s, $a = +3$ m/s^2

(b) $x_0 = +48$ m, $v_0 = 0$ m/s, $a = -3$ m/s^2

(c) $x_0 = -48$ m, $v_0 = +12$ m/s, $a = -2$ m/s^2

(d) $x_0 = +48$ m, $v_0 = -12$ m/s, $a = +2$ m/s^2

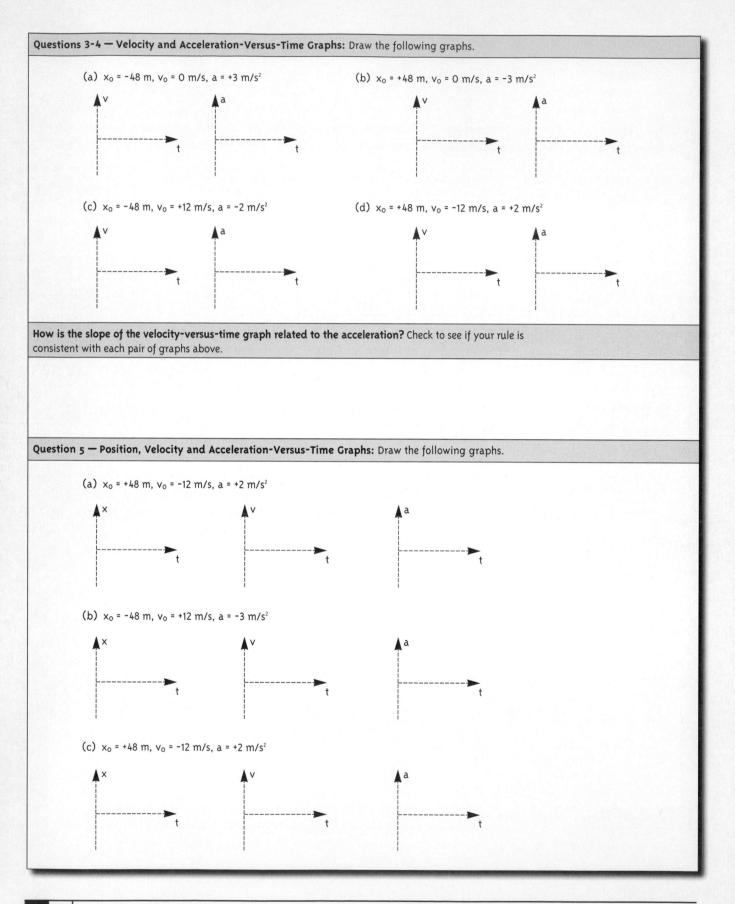

How is the slope of the velocity-versus-time graph related to the acceleration? Check to see if your rule is consistent with each pair of graphs above.

Question 5 — Position, Velocity and Acceleration-Versus-Time Graphs: Draw the following graphs.

(a) $x_0 = +48$ m, $v_0 = -12$ m/s, $a = +2$ m/s^2

(b) $x_0 = -48$ m, $v_0 = +12$ m/s, $a = -3$ m/s^2

(c) $x_0 = +48$ m, $v_0 = -12$ m/s, $a = +2$ m/s^2

In each of the following five questions, you are first given a position-versus-time graph. From the graph, you are to construct a motion diagram that is qualitatively consistent with the graph. After making the motion diagram, add velocity-versus-time and acceleration-versus-time kinematics graph lines to the position-versus-time graph. (Don't worry about the numbers for the graphs—just the general shapes.) The acceleration is constant.

Question 1:

(a) Construct a qualitative motion diagram that is consistent with the x-vs-t graph.

(b) Construct v-vs-t and a-vs-t graphs that are consistent with the x-vs-t graph.

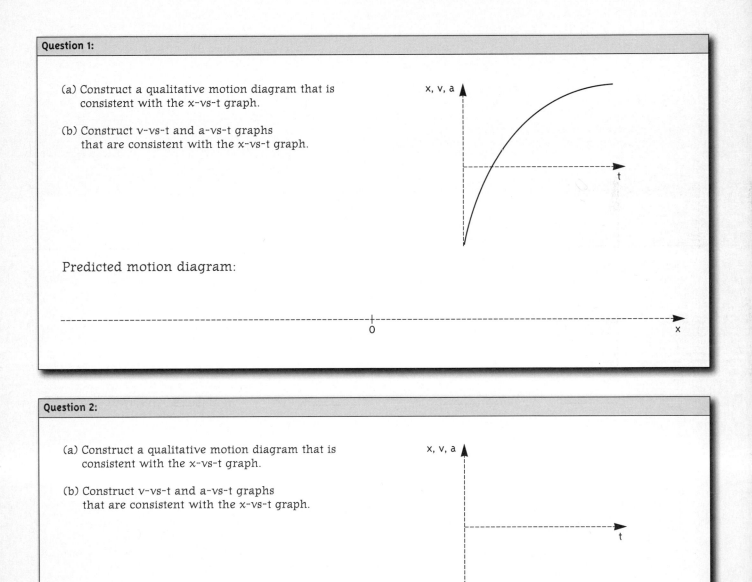

Predicted motion diagram:

Question 2:

(a) Construct a qualitative motion diagram that is consistent with the x-vs-t graph.

(b) Construct v-vs-t and a-vs-t graphs that are consistent with the x-vs-t graph.

Predicted motion diagram:

Question 3:

(a) Construct a qualitative motion diagram that is consistent with the x-vs-t graph.

(b) Construct v-vs-t and a-vs-t graphs that are consistent with the x-vs-t graph.

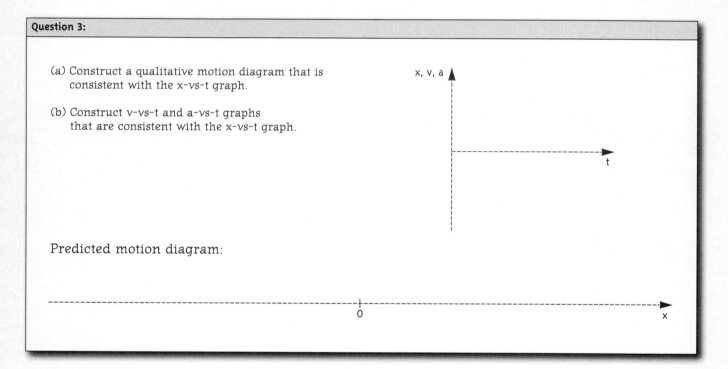

Predicted motion diagram:

Question 4:

(a) Construct a qualitative motion diagram that is consistent with the x-vs-t graph.

(b) Construct v-vs-t and a-vs-t graphs that are consistent with the x-vs-t graph.

Predicted motion diagram:

Question 5:

(a) Construct a qualitative motion diagram that is consistent with the x-vs-t graph.

(b) Construct v-vs-t and a-vs-t graphs that are consistent with the x-vs-t graph. For your motion diagram, place the dots above the horizontal axis if the object is moving right and below if moving left.

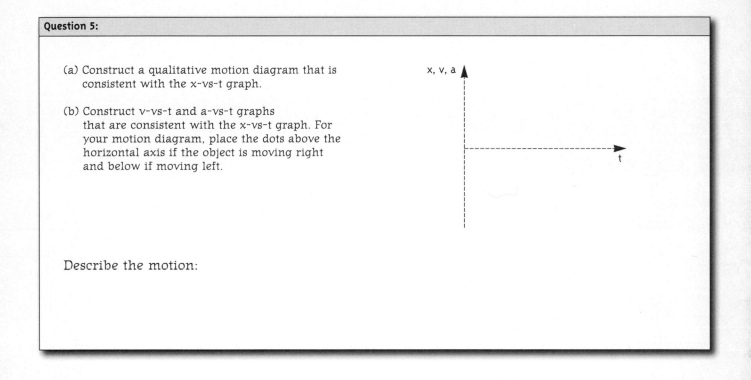

Describe the motion:

Kinematics equations describe motion. The equation used in this activity

$$x = x_0 + v_0 t + (1/2) a t^2$$

describes the changing position of an object moving along a straight line with constant acceleration. The questions test your ability to identify the values of x_0, v_0, and a.

Question 1: Run the simulation. You see a car, an equation and a white dot whose motion is described by the equation. Adjust the sliders to make the car move with the same motion as the dot — on the first try.

$$x = + 12.0 \text{ m} + (- 12.0 \text{ m/s}) + 0.5 (+ 2.0 \text{ m/s}^2) t^2$$

0   x

Question 2: Run the simulation. You see an equation and a white dot whose motion is described by the equation. Adjust the sliders to make the car move with the same motion as the dot — on the first try.

Question 3: Run the simulation. You see an equation and a white dot whose motion is described by the equation. Adjust the sliders to make the car move with the same motion as the dot — on the first try.

Question 4: Run the simulation. You see an equation and a white dot whose motion is described by the equation. Adjust the sliders to make the car move with the same motion as the dot — on the first try.

A car initially travels west at 20 m/s (about 45 mph). When the car reaches position 60 m, the brakes are applied and the car's speed decreases at a constant rate of 5.0 m/s^2 until the car stops. Describe the process using a pictorial description, a motion diagram, kinematics graphs, and kinematics position and velocity equations. When you are finished, use these descriptions to determine when and where the car stops.

Question 2 — Pictorial Description: Include a sketch, coordinate axis, symbols, and known values.

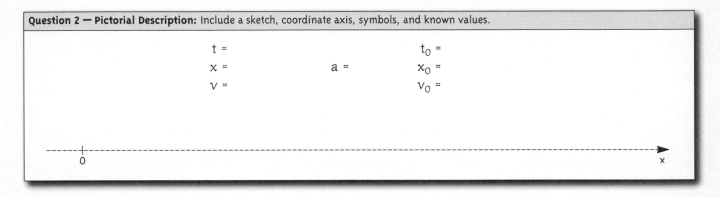

$t =$ $t_0 =$
$x =$ $a =$ $x_0 =$
$v =$ $v_0 =$

0 x

Question 3 — Motion Diagram: Draw a motion diagram. Compare arrow directions to the signs of quantities in your pictorial description.

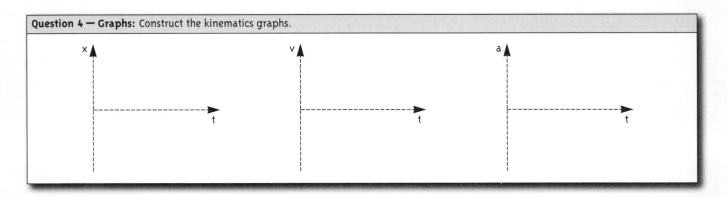

0 x

Question 4 — Graphs: Construct the kinematics graphs.

x v a

t t t

Question 5 — Equations: Use x and v equations to solve for the answer.

A skier travels 200 m to a finish line, a pole at the last tree. She starts at rest and her speed at the finish line is 31.7 m/s. Describe the process using a pictorial description, a motion diagram, kinematics graphs, and equations. Then determine the time interval needed for the trip and her constant acceleration. (Complete the descriptions below to answer Question 1.)

Pictorial Description: Include a sketch, axis, symbols, and values.

Motion Diagram: Compare directions of arrows to the signs of quantities in your pictorial description.

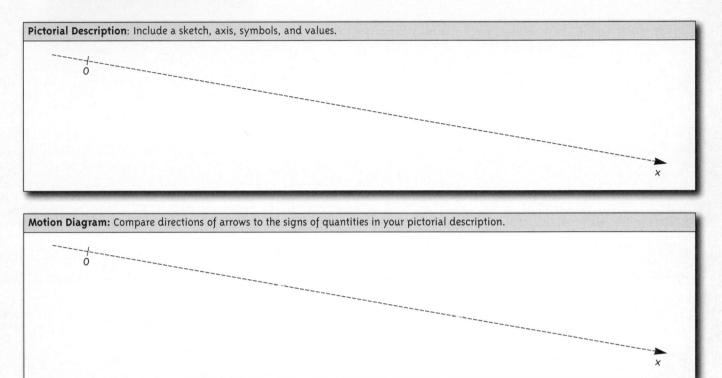

Graphs: Construct the graphs.   **Equations:** Solve for the answer.

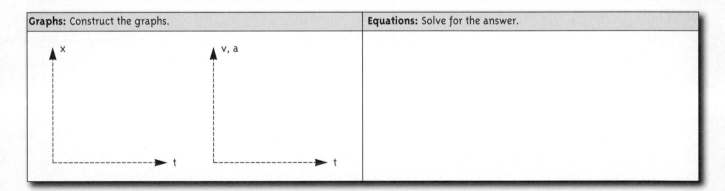

• Move the slider so that the time reads 10.1 s. Note the skier's speed at that time: _____. Also note the skier's acceleration: _____.
Relying only on the meaning of acceleration and without using any equations, predict her speed at time 12.7 s.

• After moving the slider to 12.7 s, predict the skier's velocity at 3.0 s.
Check your prediction by moving the slider back so the meter reads 3.0 s.

A balloonist ascending at a constant speed of 10 m/s accidentally releases a cup of lemonade when 15 m above the head of a crew person directly below the balloon. Determine the time interval that the crew person has to dodge the lemonade. Assume that the gravitational constant is 10 m/s^2.

Question 1 — Pictorial Description:

y

$t_0 =$

$y_0 =$

$v_0 =$

$a =$

$t =$

0 — $y =$

$v =$

Question 4 — Motion Diagram:

y

0 —

Question 5 — Graphs:

x

t

v, a

t

Question 6 — Equations and Solution

Question 1: A truck traveling at 10 m/s (about 22 mph) runs into a very thick bush and stops uniformly in a distance of 1.0 m. Determine the average acceleration of the truck during the collision.

(a) Pictorial description:

(b) Motion diagram:

(C) Equations and solution:

Question 2: Repeat your calculation, but this time determine the acceleration if the initial speed is 20 m/s. After completing your work, adjust the speed slider in the simulation to 20 m/s and check your answer.

Question 3: You doubled the initial speed from 10 m/s to 20 m/s. Qualitatively, why did the acceleration quadruple instead of double, assuming the same stopping distance?

Question 4: Why wear seat belts? The crate on the flat bed of the truck simulates a person wearing no seat belt. Observe very carefully the acceleration of the crate when it hits the hard surface at the back of the truck's cab. Based on the maximum acceleration of the crate, estimate its stopping distance.

1.8 Estimation Video: Egg Hits Windshield

Based on your observations of the video at the end of this activity, estimate acceleration of the egg while stopping when (a) belted into it's seat and (b) when unbelted.

Concept(s) to be used:

Known or estimated quantities:

Unknown to be determined:

Calculations:

You are asked to help the state motor-vehicle department construct a table that gives the car-stopping distance for different initial car speeds. Suppose that a car's initial speed is 24 m/s (about 54 mph) and that the car's acceleration when the brakes are applied is -6.0 m/s^2. There is a 0.80-s reaction time from the instant the driver sees the need to stop until the brake is applied and acceleration starts. Your goal is to predict the car's stopping distance. Note: This is a two-part problem. (Complete the descriptions below to answer Questions 1–3.)

Pictorial Description:

0   x

Motion Diagram:

0   x

Equations and Solution:

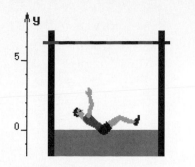

The problem starts with a pole-vaulter at the peak of his vault 6.1 m above the surface of a cushion. His fall stops after he sinks 0.40 m into the cushion. Determine the acceleration (assumed constant) that the vaulter experiences while he sinks into the cushion. Assume that the gravitational constant is 10 m/s^2. (Complete the descriptions below to answer Questions 1–5.)

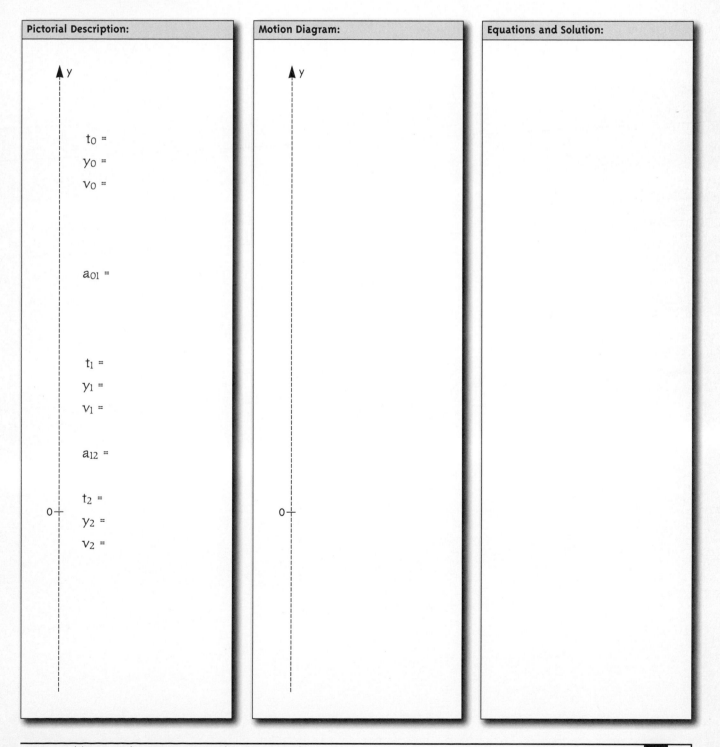

Pictorial Description:

y

$t_0 =$

$y_0 =$

$v_0 =$

$a_{01} =$

$t_1 =$

$y_1 =$

$v_1 =$

$a_{12} =$

$t_2 =$

$y_2 =$

$v_2 =$

0

Motion Diagram:

y

0

Equations and Solution:

A car starts at rest and accelerates at 4.0 m/s^2 until it reaches a speed of 20m/s. The car then travels for an unknown time interval at a constant speed of 20 m/s. Finally, the car decelerates at 4.0 m/s^2 until it stops. The car travels a total distance of 300 m. Determine the time interval needed for the entire trip. (Complete the questions below to answer Questions 1–5.)

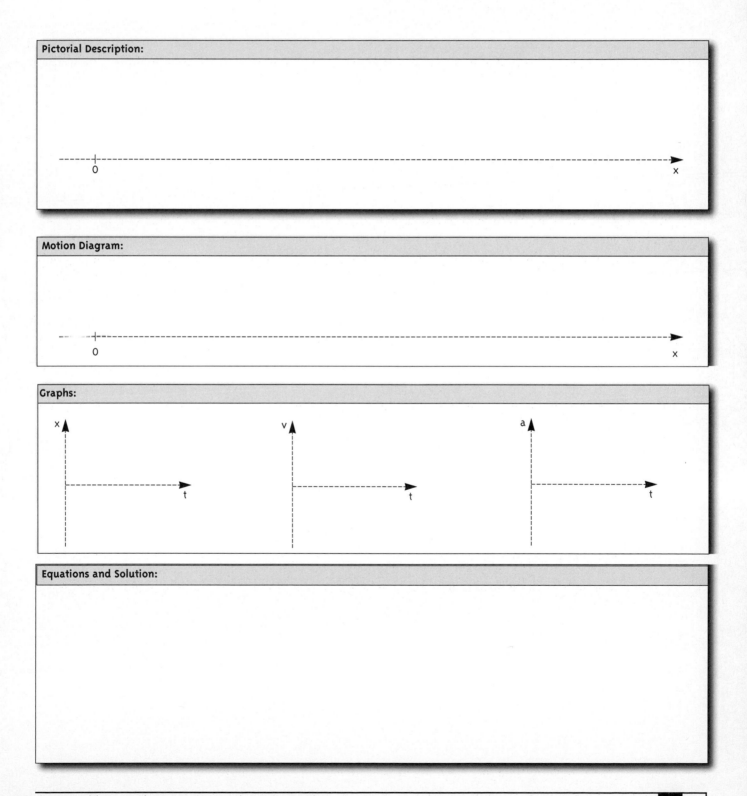

Pictorial Description:

0   x

Motion Diagram:

0   x

Graphs:

x   t   v   t   a   t

Equations and Solution:

The green car is traveling east at a constant 6.8 m/s speed. The front of the green car is initially at position zero. At the same instant, the front of the blue car is at position 12 m. The blue car is traveling west at 5.9 m/s. The blue car's speed decreases at a rate of (2.0 m/s)/s = 2.0 m/s². For each vehicle, construct a pictorial description, a motion diagram, kinematics graphs, and equations that describe the process. Then use these descriptions to determine the time when the fronts of the two vehicles meet, and the position where they meet.

Question 1 — Pictorial Description:

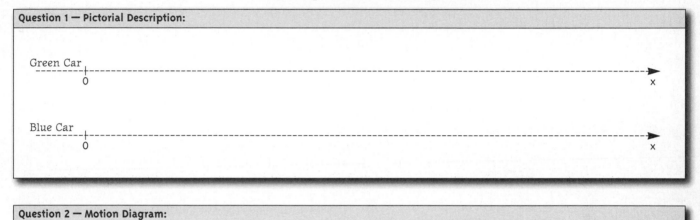

Question 2 — Motion Diagram:

Question 3 — Graphs: Show both cars on the same graph.

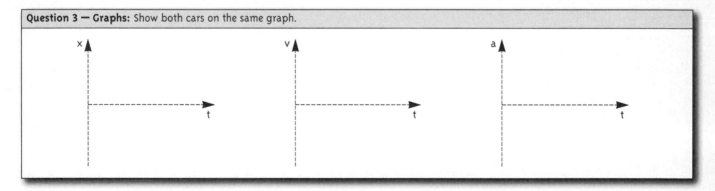

Questions 4-5 — Equations and Solution:

Green Car   Blue Car

A car at rest at a traffic light starts to move forward at the instant that a truck moving at a constant speed of 12.0 m/s passes. The car accelerates at 4.0 m/s^2. At what time and at what position does the car catch the truck? For each vehicle, construct a pictorial description, a motion diagram, kinematics graphs, and equations that describe the process. Then use these descriptions to solve the problem.

Question 1 — Pictorial Description:

Truck
0 ---→ x

Car
0 ---→ x

Question 2 — Motion Diagram:

Truck
0 ---→ x

Car
0 ---→ x

Question 3 — Graphs: Show the truck and car on the same graph.

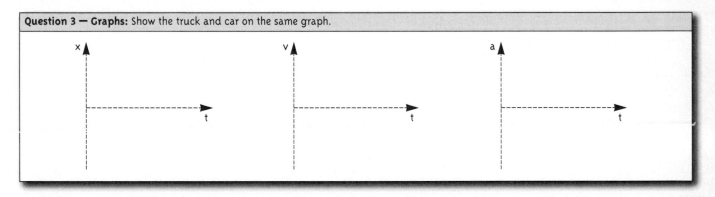

Questions 4-5 — Equations and Solution:

Truck   Car

You are hired by a movie studio to choose the parameters for a movie scene involving a car and a truck. At the start, the driver in the moving car hits the brakes. The truck starts at rest, and it's back end is initially 15.2 m ahead of the front of the car. The truck driver hits the accelerator. You are to decide the acceleration of the car and truck and the car's initial velocity so that farther down the road, the front of the car just barely misses the back of the truck. After finding slider values in Question 1 that just avoid the collision, develop the physics theory for the movie studio that supports your observations.

Note: More than one set of slider values will do the job.

Hint: Two conditions are necessary in order to avoid the collision.

Question 2 — Pictorial Description:

Truck
0 x

Car
0 x

Question 3 — Motion Diagram:

Truck
0 x

Car
0 x

Question 5 — Equations and Solution:

Truck   Car

2

FORCES AND MOTION

Constructing Free-Body Diagrams

Draw a sketch of the situation described
in the problem

Use a line to encircle and identify the system,
(the object(s) of interest)

System

Environment
touches
system here

Look along the system boundary for objects in
the environment that touch objects in the
system. Choose symbols for the forces caused by
these touching objects. Also, represent in
symbol form any long-range forces exerted on
the system. Describe in words the environmental
object causing each force and the part of the
system on which the force is exerted.

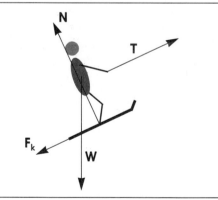

Short-range forces

Long-range force

N and **F$_k$**: Normal and kinetic friction forces caused by the snow on the skis.

T: Tension force caused by the rope on the skier.

W: Weight force caused by the gravitational pull of the earth on the person

---

Draw a separate sketch of the object(s) in the
system. Then, draw arrows representing all
forces exerted on the system. Label the arrows
with the same symbols as used above. If
possible, try to make the lengths of the arrows
representative of the relative magnitudes of
the forces.

N

T

F$_k$

W

Pretend that the system has become a point
particle and move the force arrows to the
origin of a set of coordinate axes. Make one axis
parallel to the direction of motion and the other
axis perpendicular. The head of a coordinate
axis arrow points in the positive direction.
Do not use two-headed coordinate axis arrows!

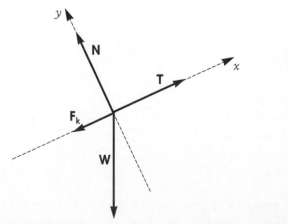

y

N

T

x

F$_k$

W

Comparing Force Magnitudes

A cable pulls the crate at constant velocity across a horizontal surface. Compare the magnitudes of the rope tension force T and the resistive kinetic friction force F_k.

(a) $T > F_k$ (b) $T = F_k$ (c) $T < F_k$

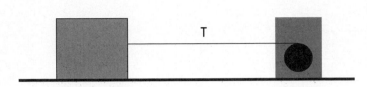

Question 1: Explain your choice.

Skydiver

The skydiver, after accelerating and falling for a short time, continues to fall at a constant terminal velocity. Compare the magnitudes of the downward weight force w and the resistive drag force of the air F_d.

(a) $w > F_d$ (b) $w = F_d$ (c) $w < F_d$

Constant v

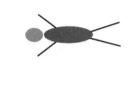

Question 1: Explain your choice.

Tension Change

The crate is pulled across a horizontal frictionless surface by a cable with tension 2.5 N. Part way along the trip, the tension is quickly reduced to 1.25 N. Complete the qualitative velocity-versus-time and acceleration-versus-time graphs, starting at the instant the tension is reduced.

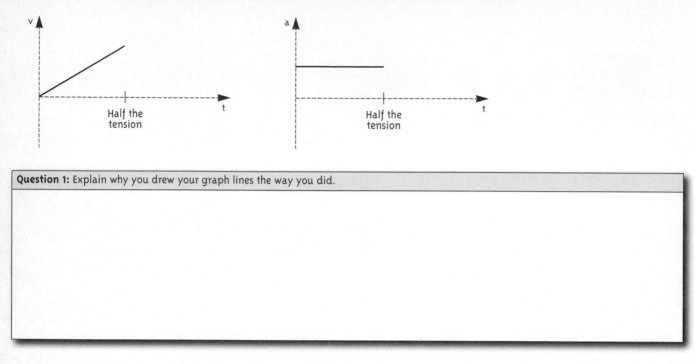

Question 1: Explain why you drew your graph lines the way you did.

Sliding on an Incline

The graphs show the velocity and the acceleration of a block that has been sliding upward along an incline. There is a small friction force exerted by the incline on the block. Complete the graph lines for the return trip back down the incline.

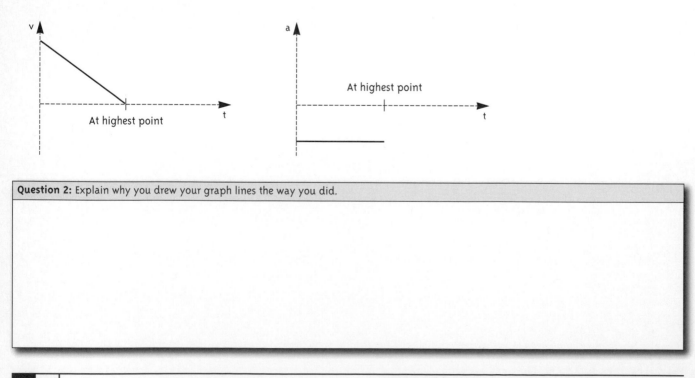

Question 2: Explain why you drew your graph lines the way you did.

Car Race

The two cars start with the same speed. The top car moves across a frictionless level surface. The bottom car moves down a frictionless incline and then back up again to the same level at which it started. Which car arrives at the finish line first, or do they tie?

(a) Car on level surface arrives first.   (b) Car on curved surface arrives first.   (c) It's a tie.

Question 1: Explain your answer.

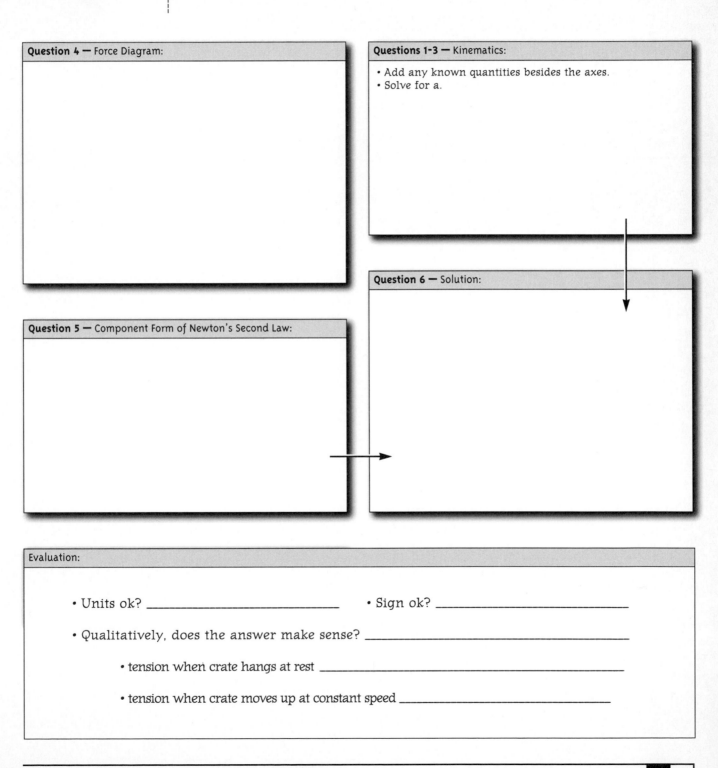

A 60-kg crate attached to a rope is initially moving upward at speed 4.0 m/s. The crate's speed decreases uniformly until it stops after traveling 4.0 m. Determine the tension in the rope while the crate's speed is decreasing. The gravitational constant is 10 N/kg.

Question 4 — Force Diagram:

Questions 1-3 — Kinematics:

• Add any known quantities besides the axes.
• Solve for a.

Question 5 — Component Form of Newton's Second Law:

Question 6 — Solution:

Evaluation:

• Units ok? _____   • Sign ok? _____

• Qualitatively, does the answer make sense? _____

   • tension when crate hangs at rest _____

   • tension when crate moves up at constant speed _____

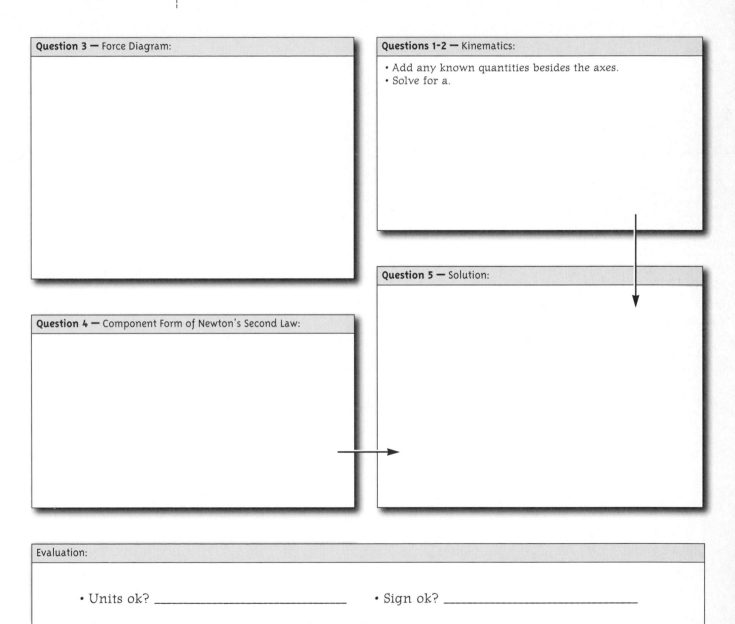

A 60-kg crate attached to a rope is initially moving downward at speed 4.0 m/s. The rope causes the crate's speed to decrease uniformly to a stop in 2.5 s. Determine the tension in the rope while the crate's speed is decreasing. The gravitational constant is 10 N/kg.

Question 3 — Force Diagram:

Questions 1-2 — Kinematics:

• Add any known quantities besides the axes.
• Solve for a.

Question 4 — Component Form of Newton's Second Law:

Question 5 — Solution:

Evaluation:

• Units ok? _____   • Sign ok? _____

• Qualitatively, does the answer make sense? _____

    • tension when crate hangs at rest _____

    • tension when crate moves down at constant speed _____

The following two pages are used to analyze the vertical motion of a toy rocket. The rocket starts at rest. Fuel emitted from the engine produces an upward force on the engine and in turn on the rocket. This upward force is called the thrust force.

Sliders allow you to change

- the magnitude of this upward thrust force
- the time interval that the thrust is exerted
- the rocket's initial upward speed

Meters indicate

- the position of the top of the rocket
- the velocity of the rocket
- the acceleration of the rocket
- the time (with $t_0 = 0$ being the time the thrust starts)

Question 1

A 5.0-N thrust is exerted on the 0.50-kg rocket for a time interval of 8.0 s. Determine the height of the rocket at the end of that time interval. The gravitational constant is 10 N/kg.

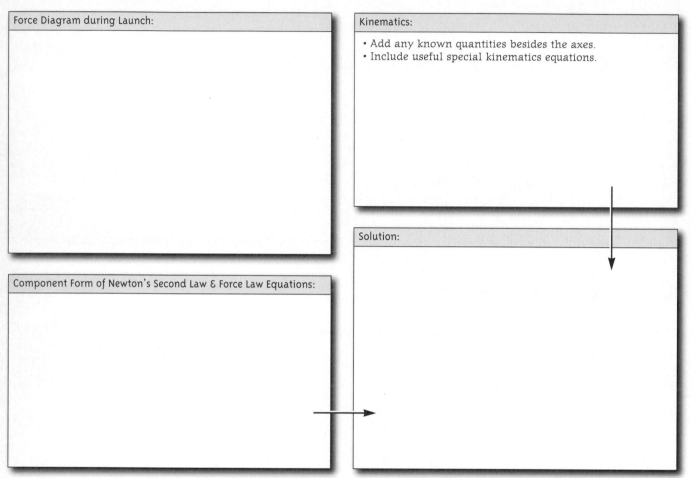

Force Diagram during Launch:

Kinematics:

- Add any known quantities besides the axes.
- Include useful special kinematics equations.

Solution:

Component Form of Newton's Second Law & Force Law Equations:

Question 2

This is the same situation in terms of thrust and rocket weight as in Question 1. Now, the rocket is initially moving up at 30 m/s. Describe (in words) what will happen to the rocket.

Questions 3-4

A 20.0-N thrust is exerted on the 0.50-kg rocket for a time interval of 8.0 s. The rocket then coasts upward more distance until its speed is reduced to zero. Determine the maximum height of the rocket. The gravitational constant is 10 N/kg.

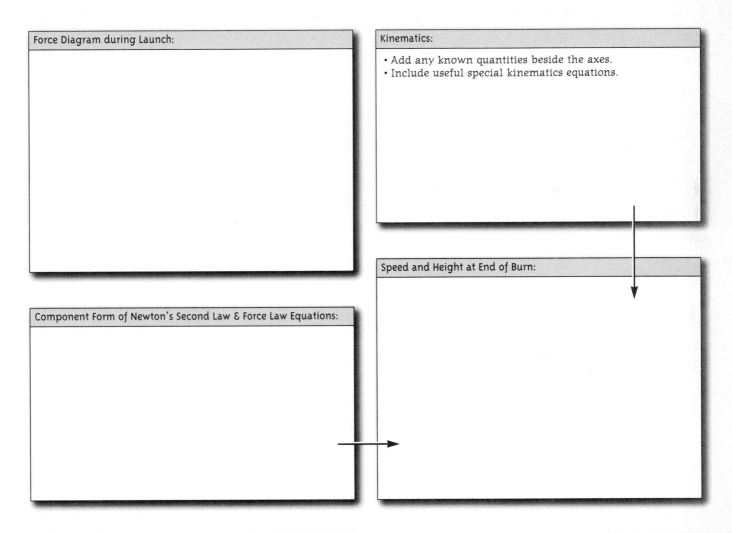

Force Diagram during Launch:

Kinematics:

- Add any known quantities beside the axes.
- Include useful special kinematics equations.

Speed and Height at End of Burn:

Component Form of Newton's Second Law & Force Law Equations:

After Burn:

Determine the time of flight after the burn and the additional distance traveled. Add this to the height at the end of the burn to find the total distance traveled.

A rope connecting a truck and a crate makes a 21.68° angle with the horizontal. The rope drags the crate along a horizontal surface. The gravitational constant is 10 N/kg. Adjust the sliders on the simulation as follows:

- crate mass = 50 kg
- coefficient of friction = 0.60
- acceleration (in the x-direction of the truck and the crate) = +3.0 m/s²

Determine if the meter readings are consistent with the y-component form of Newton's Second Law and with the kinematics equations of motion.

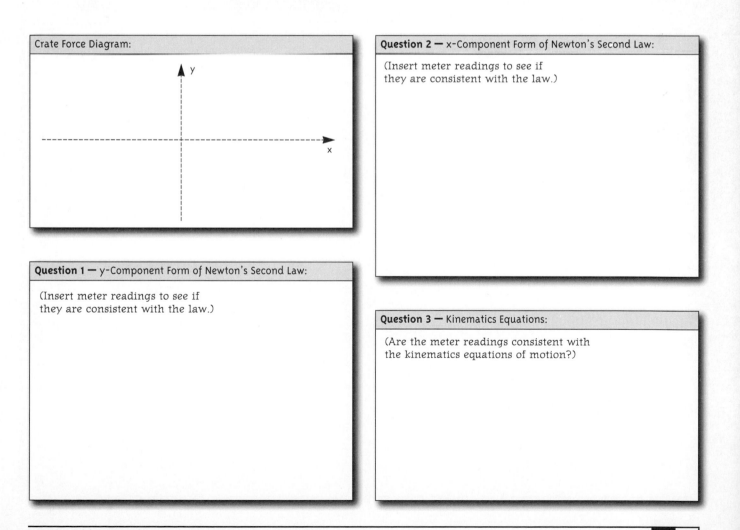

Crate Force Diagram:

y

x

Question 2 — x-Component Form of Newton's Second Law:

(Insert meter readings to see if they are consistent with the law.)

Question 1 — y-Component Form of Newton's Second Law:

(Insert meter readings to see if they are consistent with the law.)

Question 3 — Kinematics Equations:

(Are the meter readings consistent with the kinematics equations of motion?)

(a) Estimate the mass of the green block. The coefficient of kinetic friction between the block and the horizontal surface is 0.2.

(b) Why is the tension in the string so small in the second movie compared to the tension in the first movie?

Concept(s) to be used:

Known or estimated quantities:

Unknown to be determined:

Calculations:

You exert a constant 10-N force directed 37° below the horizontal while pushing a 1.0-kg crate along a level surface. The coefficient of friction between the surface and the crate is 0.20, and the gravitational constant is 10 N/kg. Determine the magnitudes of the crate's weight, the normal force, the kinetic friction force, and the acceleration of the crate.

Question 1 — Force Diagram:

y

x

Questions 2-3 — y-Component Form of Newton's Second Law & Interaction Equations:

Question 4 — Normal Force:

Kinetic Friction Force:

Question 5 — Acceleration:

You exert a constant 20-N force directed 53° above the horizontal while pushing a 1.0-kg crate up a vertical wall. The coefficient of friction between the surface and the crate is 0.20, and the gravitational constant is 10 N/kg. Determine the magnitudes of the crate's weight, the normal force, the kinetic friction force, and the acceleration of the crate.

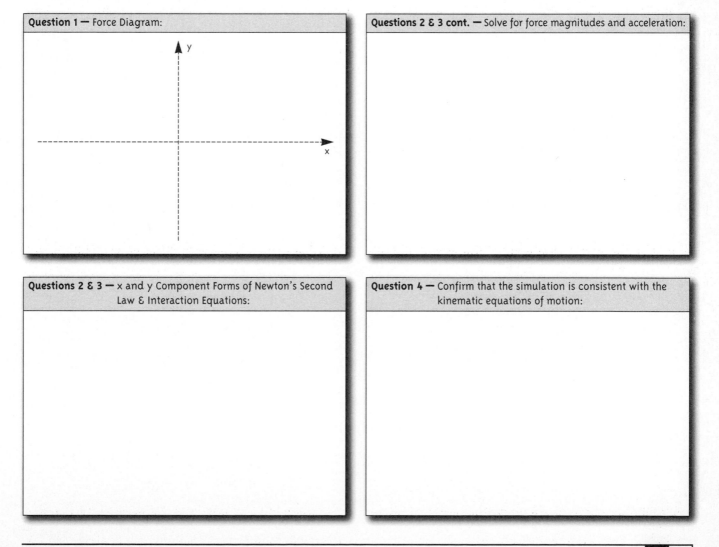

Question 1 — Force Diagram:

Questions 2 & 3 cont. — Solve for force magnitudes and acceleration:

Questions 2 & 3 — x and y Component Forms of Newton's Second Law & Interaction Equations:

Question 4 — Confirm that the simulation is consistent with the kinematic equations of motion:

The 100-kg skier skis down a steep ski slope which is inclined at 26° to the horizontal. The coefficient of friction between the skis and the snow is a sticky 0.30, and the gravitational constant is 10 N/kg. Determine the magnitudes of the weight, the normal force, the kinetic friction force, the acceleration, and the speed after traveling 200 m. The skier starts at rest.

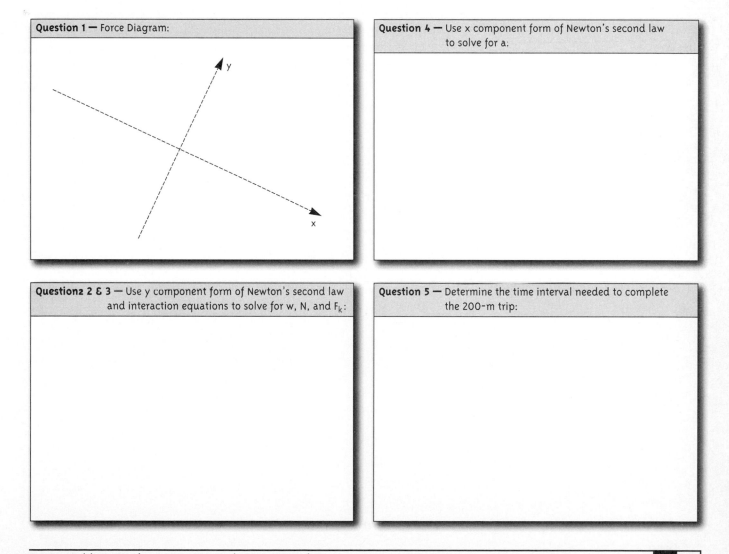

Question 1 — Force Diagram:

Question 4 — Use x component form of Newton's second law to solve for a:

Question2 2 & 3 — Use y component form of Newton's second law and interaction equations to solve for w, N, and F_k:

Question 5 — Determine the time interval needed to complete the 200-m trip:

Estimate the coefficient of friction between the wheels of the truck and the inclined surface. Indicate if the friction is kinetic friction or static friction.

Concept(s) to be used:

Known or estimated quantities:

Unknown to be determined:

Calculations:

A rope pulls the 100-kg skier up a steep slope inclined at 38.5° to the horizontal. The coefficient of friction between the skis and snow is a sticky 0.20, and the gravitational constant is 10 N/kg. Determine the magnitude of the rope tension needed to pull the skier at a constant 1.0 m/s speed.

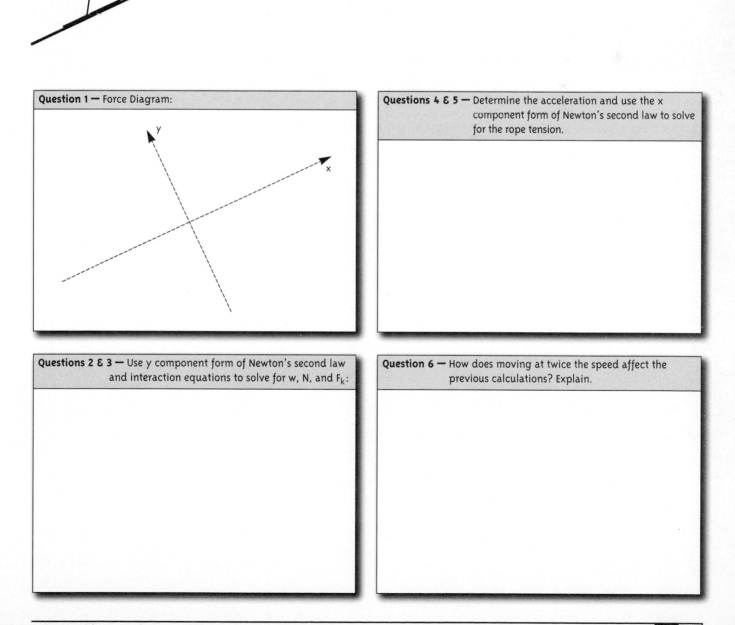

Question 1 — Force Diagram:

Questions 4 & 5 — Determine the acceleration and use the x component form of Newton's second law to solve for the rope tension.

Questions 2 & 3 — Use y component form of Newton's second law and interaction equations to solve for w, N, and F_k:

Question 6 — How does moving at twice the speed affect the previous calculations? Explain.

You exert a constant 30-N horizontal force while pushing a 2.0-kg crate up a steep 37° incline. (Note that the force is not parallel to the incline but horizontal.) The coefficient of friction between the surface and the crate is 0.143, and the gravitational constant is 10 N/kg. Determine the magnitudes of the crate's weight, the normal force, the kinetic friction force, and the acceleration of the crate.

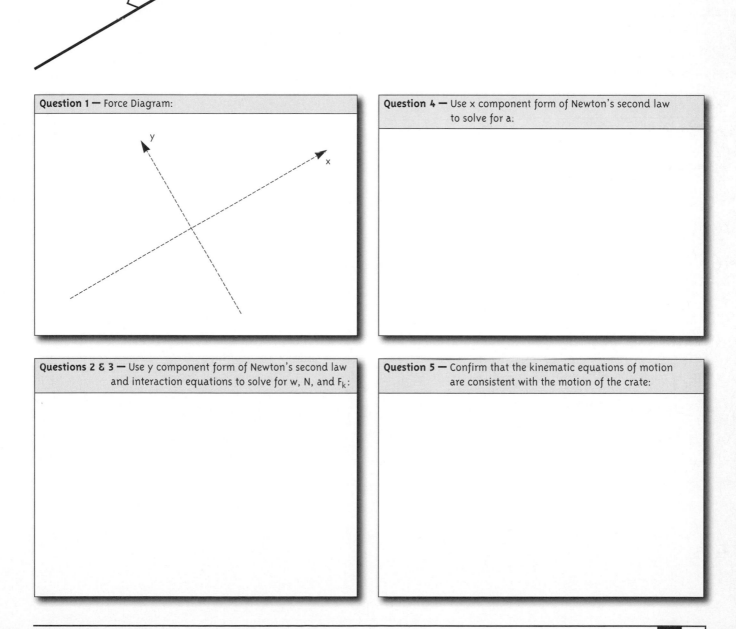

Question 1 — Force Diagram:

Question 4 — Use x component form of Newton's second law to solve for a:

Questions 2 & 3 — Use y component form of Newton's second law and interaction equations to solve for w, N, and F_k:

Question 5 — Confirm that the kinematic equations of motion are consistent with the motion of the crate:

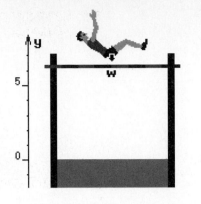

The 70-kg pole-vaulter at the top of his flight has zero speed and is 6.1 m above the cushion below. After he falls, he sinks 0.4 m into the cushion while stopping. Assuming uniform acceleration, determine the average force of the cushion on the vaulter while the cushion is stopping him. The gravitational constant is 10 N/kg. (Complete the information below to answer Questions 1–5.)

Plan a Solution.

Complete the solution.

Be sure to include a force diagram and apply Newton's Second Law. (Include ALL forces shown in the diagram.)

A 200-N force pushes forward on the 20-kg truck. The truck pulls two crates connected together by ropes, as shown. The gravitational constant is 10 N/kg. Determine the acceleration of the crates and truck and the tension in each rope for one of the following sets of conditions.

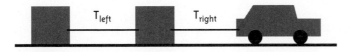

	Set 1 Question 1	Set 2 Question 4
Mass of Left Crate	20 kg	20 kg
Mass of Right Crate	20 kg	20 kg
Friction Coefficient	0.00	0.40

Left Crate	Right Crate	Truck
Force Diagram:	Force Diagram:	Force Diagram:
x-Component Form of Newton's Second Law:	x-Component Form of Newton's Second Law:	x-Component Form of Newton's Second Law:

Combine the above to solve for the unknowns.

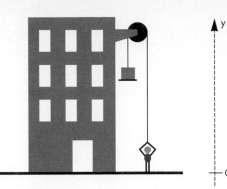

The 44-kg bricklayer plans to lower the barrel of bricks from the fourth floor of the building. Unfortunately, the mass of the bricks is 56 kg, and the bricklayer is pulled up in the air. Determine the speed of the bricklayer after traveling 7.0 m upward — the distance that the bricks fall. The gravitational constant is 10 N/kg.

Question 1 — Force Diagram for Descending Bricks:	**Question 3 —** Force Diagram for Ascending Bricklayer:

Question 2 — Component Form of Newton's Second Law:	**Question 4 —** Component Form of Newton's Second Law:

Write an equation to relate the accelerations of the two objects. Determine a and T.

Question 5 — Use one of the kinematics equations to solve for v.

Determine the rope tension and the acceleration of the system with the hanging grey block's mass equal to 5.0 kg and the sliding red block's mass equal to 5.0 kg. Ignore friction and assume that the gravitational constant is 10 N/kg. (Complete the information below to answer Questions 1–6.)

Force Diagram for Block on Table:	Force Diagram for Descending Block:
Component Form of Newton's Second Law:	Component Form of Newton's Second Law:

Determine a and T.

Estimate the coefficient of friction between the wheels of the truck and the inclined surface. The truck's mass is 795 g and the hanging blocks' combined mass is 300 g. Be sure to justify your estimate based on physics reasoning.

Concept(s) to be used:

Unknown to be determined:

Known or estimated quantities:

Calculations:

3

PROJECTILE MOTION

If we ignore air resistance, the motion of a projectile can be considered as the combination of two independent types of motion, each described by its own set of equations.

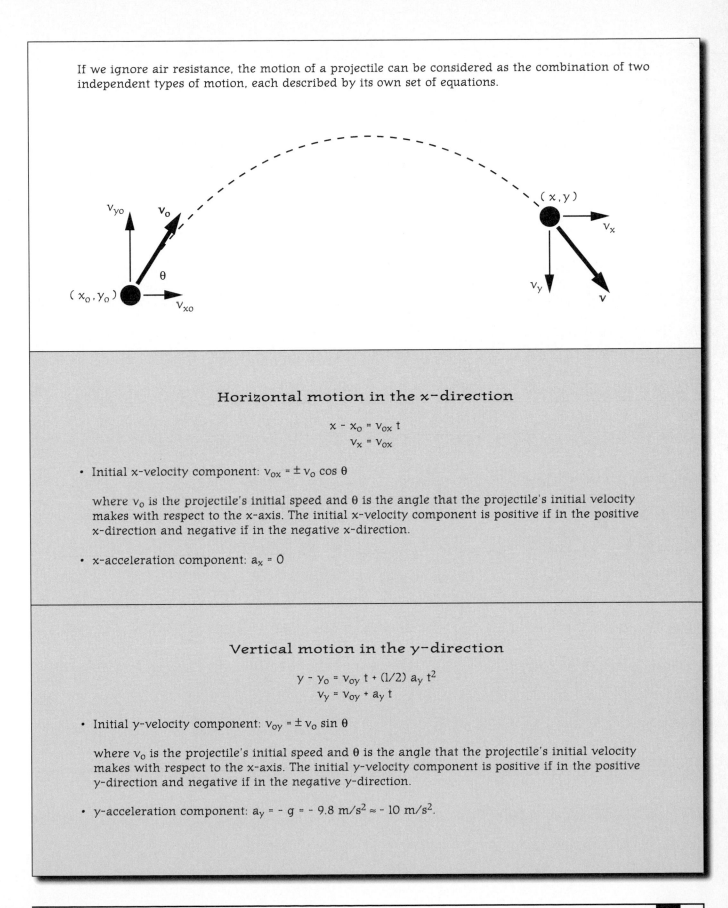

Horizontal motion in the x-direction

$$x - x_o = v_{ox} \, t$$
$$v_x = v_{ox}$$

- Initial x-velocity component: $v_{ox} = \pm v_o \cos \theta$

  where v_o is the projectile's initial speed and θ is the angle that the projectile's initial velocity makes with respect to the x-axis. The initial x-velocity component is positive if in the positive x-direction and negative if in the negative x-direction.

- x-acceleration component: $a_x = 0$

Vertical motion in the y-direction

$$y - y_o = v_{oy} \, t + (1/2) \, a_y \, t^2$$
$$v_y = v_{oy} + a_y \, t$$

- Initial y-velocity component: $v_{oy} = \pm v_o \sin \theta$

  where v_o is the projectile's initial speed and θ is the angle that the projectile's initial velocity makes with respect to the x-axis. The initial y-velocity component is positive if in the positive y-direction and negative if in the negative y-direction.

- y-acceleration component: $a_y = -g = -9.8 \text{ m/s}^2 \approx -10 \text{ m/s}^2$.

Question 1: Two balls start 5.0 m above a brick surface, each with a zero y-velocity component. Ball A's initial x-velocity component is zero, and Ball B's initial x-velocity component is 5.0 m/s. Which ball reaches the floor first or do they tie? Explain your choice.

(a) Ball A   Explain your prediction:

(b) Ball B

(c) It's a tie.

Question 2: Try other initial x-velocity component settings for the two balls. Based on your observations, develop a rule for how the x-velocity component affects the time interval that a ball needs to fall to the floor.

Question 3: It takes 1.0 s for the balls to fall 5.0 m, assuming their initial y-velocity component is zero, that air resistance is negligible, and that the gravitational constant is 10 m/s². Choose three different pairs of initial x-velocity component settings for the two balls so that they touch just as they reach ground level, the floor.

Adjust the x-velocity component slider to 2.0 m/s. Watch the simulation very carefully and record

- the maximum height reached by the projectile _____
- the flight time _____
- the range of the projectile (the horizontal distance it travels) _____

Each division on the simulation is 1.0 m, and the gravitational constant is 10 m/s^2. Predict how each of these quantities will change when the horizontal x-velocity component is changed from 2.0 m/s to 10.0 m/s and the simulation is run again.

1. The maximum elevation: (a) increases   (b) remains the same   (c) decreases

Explain your prediction:

2. The flight time: (a) increases   (b) remains the same   (c) decreases

Explain your prediction:

3. The range of the projectile: (a) increases   (b) remains the same   (c) decreases

Explain your prediction:

After completing your predictions and the reasons for the predictions, run the simulation to see how you did. Modify your reasoning if necessary.

Question 1: Determine the x-acceleration component of the ball by calculating the change in the x-velocity component during a time interval divided by that time interval.

Question 2: Determine the y-acceleration component of the ball by calculating the change in the y-velocity component during a time interval divided by that time interval. Do the calculation for the following time intervals:

0.0 s to 1.0 s:

1.0 s to 2.0 s:

2.0 s to 3.0 s:

3.0 s to 4.0 s:

What do you think the y-acceleration component is for the entire trip?

Question 3: By how much does the magnitude of the y-velocity component change during the following time intervals?

$\Delta t = 1.0$ s:

$\Delta t = 2.0$ s:

$\Delta t = 0.10$ s:

- The x-component of the initial velocity v_{xo} is

$$v_{xo} = \pm v_o \cos \theta$$

where the plus sign is used if the x component points in the positive x-direction, and the minus sign is used if the x-component points in the negative x-direction. The angle θ is the angle that v_o makes relative to the positive or negative x-axis.

- The y-component of the initial velocity v_{yo} is

$$v_{yo} = \pm v_o \sin \theta$$

where the plus sign is used if the y component points in the positive y-direction, and the minus sign is used if the y-component points in the negative y-direction. The angle θ is the angle that v_o makes relative to the positive or negative x-axis.

Question 1: Determine the x- and y-velocity components for a projectile whose initial velocity is 50 m/s at an angle of 53° above the positive x-axis.

Question 2: Determine the x- and y-velocity components for a projectile whose initial velocity is 80 m/s at an angle of 30° below the positive x-axis.

Question 3: Determine the x- and y-velocity components for a projectile whose initial velocity is 100 m/s at an angle of 37° above the negative x-axis.

Question 4: Determine the x- and y-velocity components for a projectile whose initial velocity is 100 m/s in the positive x-direction.

Question 1: $v_{oy} = 0$

Question 2: $v_{oy} = +4.0$ m/s

Question 3: $v_{ox} = +10$ m/s

Question 4: $v_{ox} = +11.0$ m/s

For each question, you adjust one initial velocity component and calculate the value of the other velocity component so that the ball lands on the target. Each division on the simulation is 1.0 m, and the gravitational constant is 10 m/s².

Make a sketch below to indicate all information known about the initial situation and the desired final situation:

Apply the x-component kinematics equations:

Apply the y-component kinematics equations:

Solve for the unknown:

Evaluate your solution in terms of magnitude, sign, and unit:

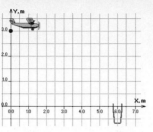

In a toy airplane contest, a small plane travels at a constant elevation and at a constant 6.0 m/s speed. You are to decide the position where the plane should release supplies so that they land in a target basket. The gravitational constant is 10 m/s².

Make a sketch below to indicate all information known about the initial situation and the desired final situation:

Apply the x-component kinematics equations:

Apply the y-component kinematics equations:

Solve for the unknown:

Evaluate your solution in terms of magnitude, sign, and unit:

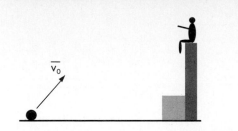

Adjust the ball's initial speed (in m/s) so that the scientist on the ledge can catch the ball. Each division on the screen is 1.0 m. The launch angle is 58.3°. Use any information provided here, distance measurements on the screen, and the projectile-motion concepts to determine the gravitational constant on this planet. HINT: For this question, you do not first use the y-axis equation to determine the projectile's flight time.

Make a sketch below to indicate all information known about the initial situation and the desired final situation:

Apply the x-component kinematics equations:

Apply the y-component kinematics equations:

Solve for the unknown:

Evaluate your solution in terms of magnitude, sign, and unit:

---

v_0

Five-foot two-inch, 109-pound Debbie Lawler, on March 31, 1973, sailed 76 feet over a line of parked cars to set the women's motorcycle distance-jumping record. Debbie's next dream is to jump her Suzuki motorcycle over two destroyers moored in the harbor. Suppose she came to ask you to be her consultant in planning the jump. To test your knowledge, she asks you if the cyclist in our simulation will make it to the opposite shore. The mass of the miniature cyclist is 20 kg. Each division on the simulation is 1.0 m, and the gravitational constant is 10 m/s². The x- and y-velocity components when the cyclist reaches the top of the ramp are 5.7 m/s and 4.2 m/s, respectively.

Make a sketch below to indicate all information known about the initial situation and the desired final situation:

Apply the x-component kinematics equations:

Apply the y-component kinematics equations:

Solve for the unknown:

Evaluate your solution in terms of magnitude, sign, and unit:

Estimate the initial speed of the shot put.
It travels 19.3 m and is in the air 1.65 s.

Concept(s) to be used:

Known or estimated quantities:

Unknown to be determined:

Calculations:

4

CIRCULAR MOTION

The direction of a moving object's acceleration when at some arbitrary position can be estimated if you know its velocity just before arriving at that position and just after. The procedure is illustrated below.

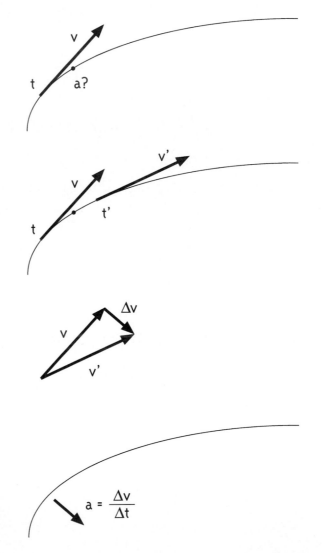

Original velocity:
Draw an arrow representing the velocity v of the object at time t just before ariving at the position of interest.

New velocity:
Draw another arrow representing the velocity v' of the object at time t' just after passing the point.

Velocity change:
To find the change in velocity Δv during the time interval $\Delta t = t' - t$, place the tails of v and v' together. The change in velocity Δv is a vector that points from the head of v to the head of v'. Notice in the figure at the right that $v + \Delta v = v'$, or by rearranging, $\Delta v = v' - v$ (that is, Δv is the change in velocity).

Acceleration:
The acceleration equals the velocity change Δv divided by the time interval Δt needed for that change: that is, $a = \Delta v / \Delta t$. If you do not know the time interval, you can at least estimate the direction of the acceleration because the acceleration arrow points in the same direction as Δv.

For each situation shown below, estimate the direction of the acceleration of the pendulum bob.

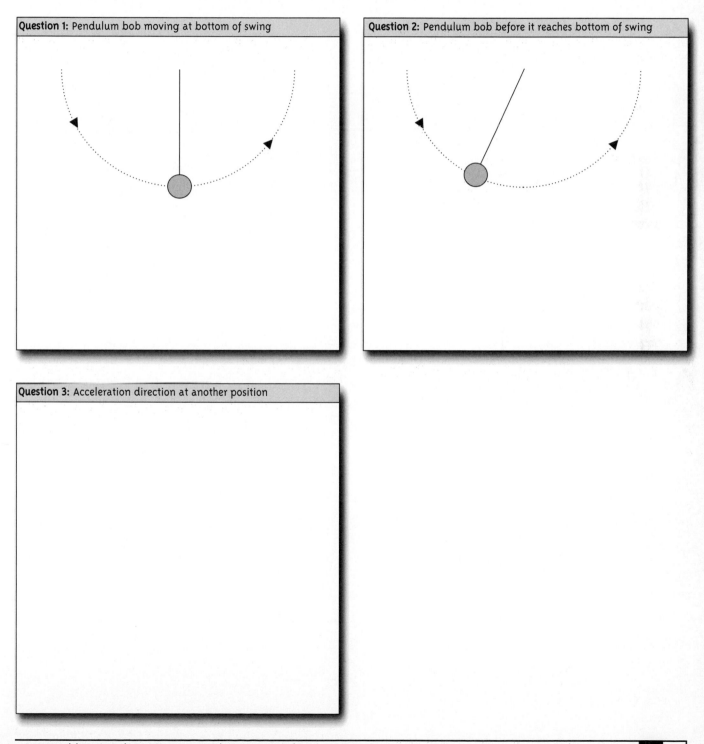

Question 1: Pendulum bob moving at bottom of swing

Question 2: Pendulum bob before it reaches bottom of swing

Question 3: Acceleration direction at another position

For Questions 1 and 2 below, estimate the direction of the acceleration of the car when the car is at the position shown in the diagram. (A top view is given.)

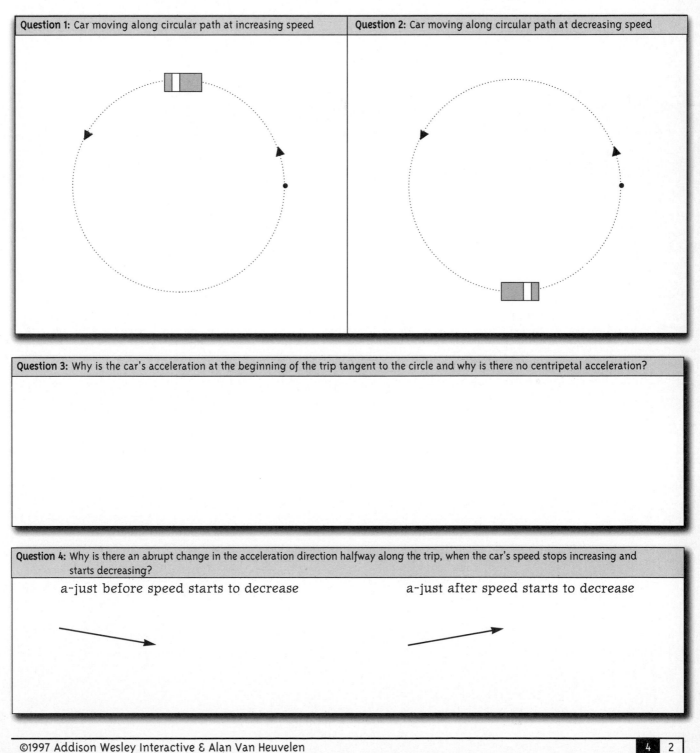

Question 1: Car moving along circular path at increasing speed

Question 2: Car moving along circular path at decreasing speed

Question 3: Why is the car's acceleration at the beginning of the trip tangent to the circle and why is there no centripetal acceleration?

Question 4: Why is there an abrupt change in the acceleration direction halfway along the trip, when the car's speed stops increasing and starts decreasing?

a-just before speed starts to decrease

a-just after speed starts to decrease

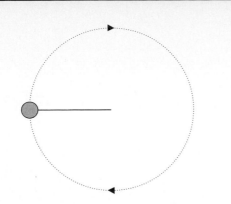

Record the meter readings for the magnitude of the centripetal acceleration as a function of speed and as a function of the radius of the circle. Then plot the data and decide how the magnitude of the acceleration depends on each quantity.

Question 1 – Dependence of Acceleration on Speed:

Speed (m/s)   Centripetal a (m/s²)

a_c

Dependence of centripetal acceleration on speed:

v

Question 2 – Dependence of Acceleration on Radius:

Radius (m)   Centripetal a (m/s²)

a_c

Dependence of centripetal acceleration on radius:

r

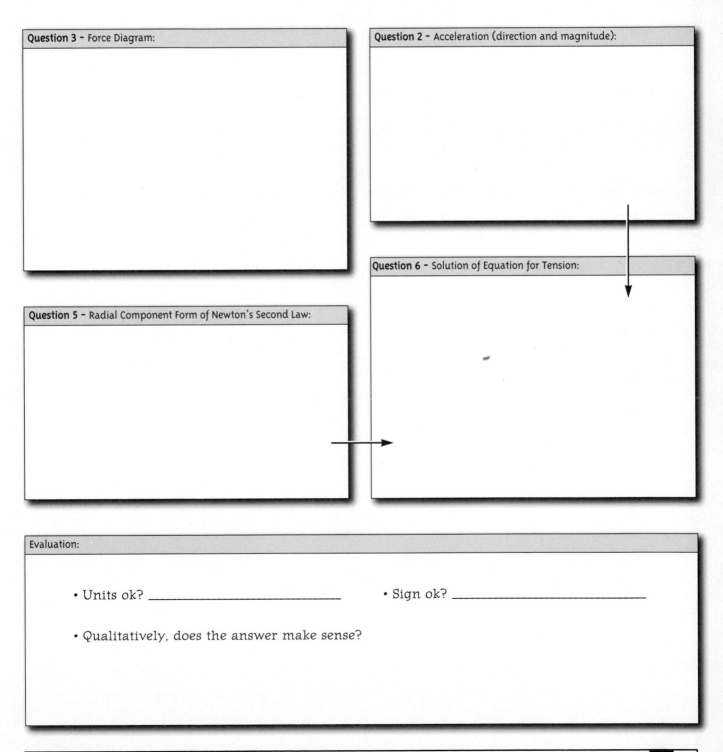

A 1.0-kg pendulum bob swings at the end of a 2.0-m-long string. Determine the tension in the string as the bob passes the lowest point in the swing. Its speed at this point is 6.2 m/s. Assume that the gravitational constant is 10 N/kg.

Question 3 – Force Diagram:

Question 2 – Acceleration (direction and magnitude):

Question 6 – Solution of Equation for Tension:

Question 5 – Radial Component Form of Newton's Second Law:

Evaluation:

• Units ok? _____   • Sign ok? _____

• Qualitatively, does the answer make sense?

4.4 Estimation Video: Swinging a Bucket of Water

Estimate the tension in the person's arm while swinging the plastic bucket that holds one gallon of water. Any other information you need will have to be estimated from the video.

Concept(s) to be used:

Unknown to be determined:

Known or estimated quantities:

Calculations:

A 10-kg cart coasts up and over a frictionless circular path of radius 2.08 m. As it passes the top of the path, its speed is 3.2 m/s. Determine the magnitude of the normal force of the hill on the cart when the cart is at the top of the hill. Assume that the gravitational constant is 10 N/kg.

Question 2 – Force Diagram:

Question 1 – Acceleration:

Question 3 – Radial Component Form of Newton's Second Law:

Question 4 – Solution for Equation for Normal Force:

Evaluation:

• Units ok? _____

• Sign ok? _____

• Qualitatively, does the answer make sense?

A 1.0-kg ball swings in a vertical circle at the end of a 4.0-m-long string. You are to determine the tension in the string when the ball is at three different locations. The speed of the ball at each location is given below. Assume that the gravitational constant is 10 N/kg.

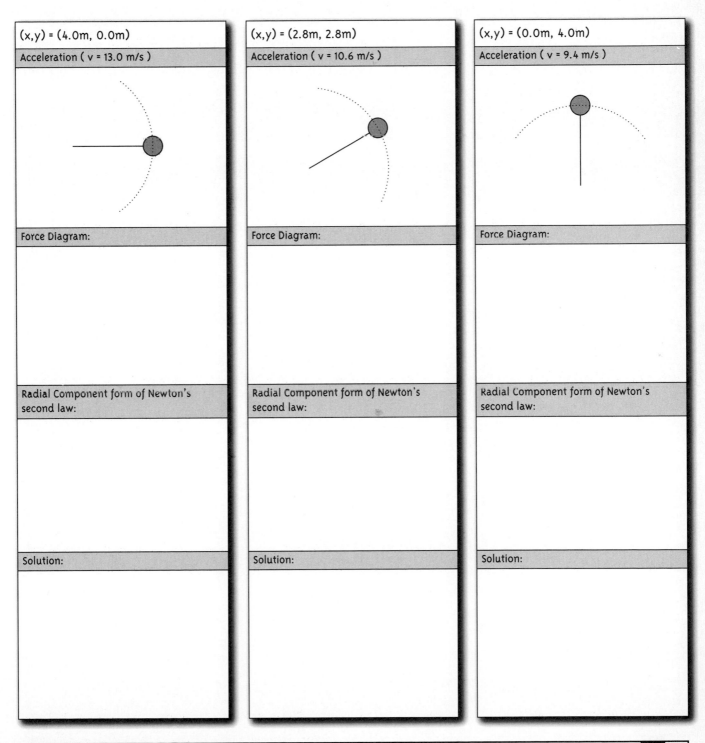

(x,y) = (4.0m, 0.0m)	(x,y) = (2.8m, 2.8m)	(x,y) = (0.0m, 4.0m)
Acceleration (v = 13.0 m/s)	Acceleration (v = 10.6 m/s)	Acceleration (v = 9.4 m/s)
Force Diagram:	Force Diagram:	Force Diagram:
Radial Component form of Newton's second law:	Radial Component form of Newton's second law:	Radial Component form of Newton's second law:
Solution:	Solution:	Solution:

A 1000-kg car moves at a maximum speed so that it does not skid off the 50-m radius level track. If the coefficient of static friction between the road and wheels is 0.80, what is the maximum speed? Assume that the gravitational constant is 10.0 N/kg = 10.0m/s.

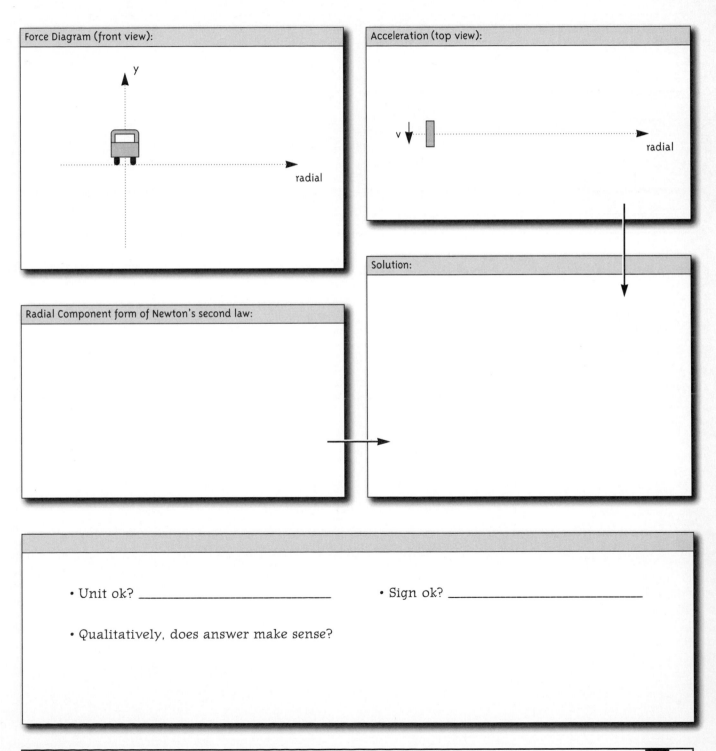

Force Diagram (front view):

y

radial

Acceleration (top view):

v

radial

Radial Component form of Newton's second law:

Solution:

• Unit ok? _____

• Sign ok? _____

• Qualitatively, does answer make sense?

Use circular dynamics (the circular form of Newton's second law) and circular kinematics to estimate the plane's speed while moving at the end of the string.

Concept(s) to be used:

Known or estimated quantities:

Unknown to be determined:

Calculations:

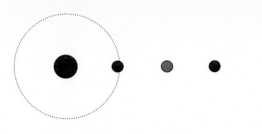

The satellite with orbital radius 1.0 unit moves in a circular orbit when its speed is 3.0 m/s. Determine the speed that the identical satellites with 2.0 and 3.0 unit radii must move to remain in circular orbits. With other speed settings, the satellites move in elliptical orbits. To solve the problem, develop a general Newton's second law theory for the satellite motion that relates speed and radius. Use the information about the 1.0 radius orbit to solve for other unknowns.

Force Diagram:

Acceleration:

Radial Component form of Newton's second law:

Satellite Equation:

Complete Problem Solution:

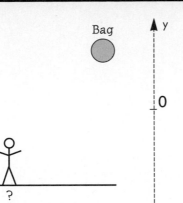

Bag

y

0

?

A 4.0-kg bag is attached to the end of a 4.0-m long rope. You swing the rope and bag in a vertical circle and release it when the bag is at the top of the swing. The tension in the rope at that point is 25.7 N. Where should a friend stand to catch the bag before it hits the ground. His head is 7.5 m below the release point of the bag. The gravitational constant is 10 N/kg.

Solution Plan:

Number of parts: _____

Indicate the conceptual knowledge used for each part and the unknown that you plan to determine for each part.

Part 1:

Force Diagram:	Radial Component form of Newton's second law:	Solve for Unknown:

Part 2:

5

WORK AND ENERGY

For each situation, determine the work done by one object on another, as specified in the question.

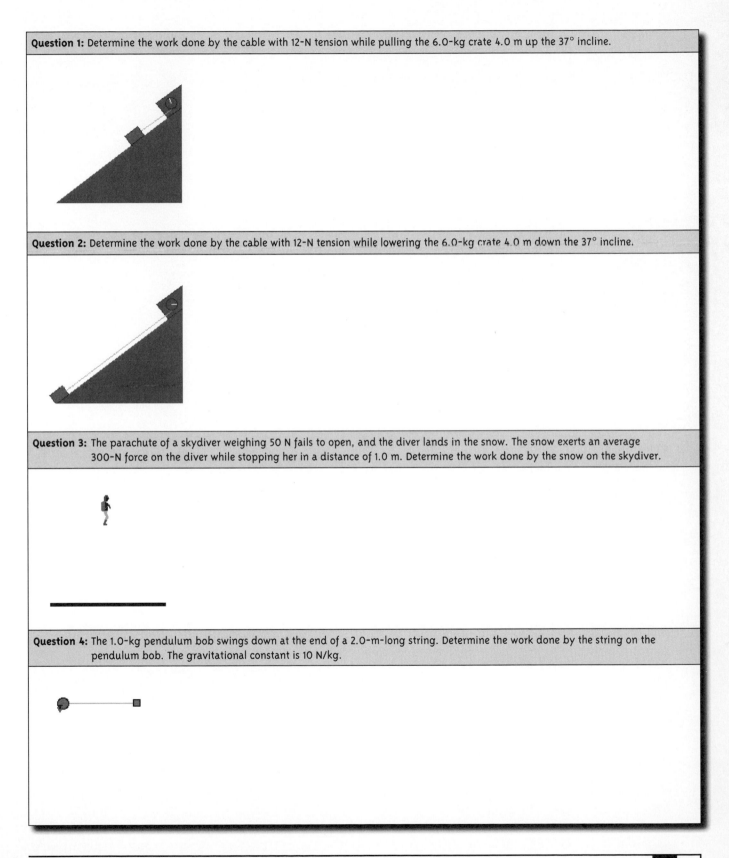

Question 1: Determine the work done by the cable with 12-N tension while pulling the 6.0-kg crate 4.0 m up the 37° incline.

Question 2: Determine the work done by the cable with 12-N tension while lowering the 6.0-kg crate 4.0 m down the 37° incline.

Question 3: The parachute of a skydiver weighing 50 N fails to open, and the diver lands in the snow. The snow exerts an average 300-N force on the diver while stopping her in a distance of 1.0 m. Determine the work done by the snow on the skydiver.

Question 4: The 1.0-kg pendulum bob swings down at the end of a 2.0-m-long string. Determine the work done by the string on the pendulum bob. The gravitational constant is 10 N/kg.

You are asked to help in the design of an ejector pad that will be used in a stunt. Your task for now is to analyze the energy transformations during the process.

Questions 1-2: Indicate the form of the energy of the earth-spring-person system at the three positions indicated in the sketch above. This is a qualitative activity, and the relative lengths of the bars are undetermined.

At rest on the pad with the spring compressed

U_{so} + U_{go} + K_o

0

At the instant the person leaves contact with the pad

U_s + U_g + K + U_{in}

0

At the person's highest position

U_s + U_g + K + U_{in}

0

Question 3: Apply the generalized work-energy equation to the process (starting at the left and ending at the right)

5.2 Estimation Video: Grasshopper Force Constant

Estimate the effective force constant of the spring-like legs of the 5.4 g grasshopper.

Concept(s) to be used:

Known or estimated quantities:

Unknown to be determined:

Calculations:

A person, initially at rest, slides down a slippery frictionless hill and then across a grassy field where friction stops the sled before it reaches a cliff on the other side.

Questions 1-2: Complete the work energy bar charts for the designated positions along the slide.

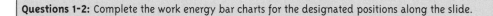

At rest on side of hill

U_{go} + U_{so} + K_o + W

0

Moving fast at bottom of hill

U_g + U_s + K + U_{in}

0

At rest after crossing field

U_g + U_s + K + U_{in}

0

For each bar chart, draw a picture of a process that is consistent with the bar chart. (There may be many processes described by the same chart.)

Question 1:

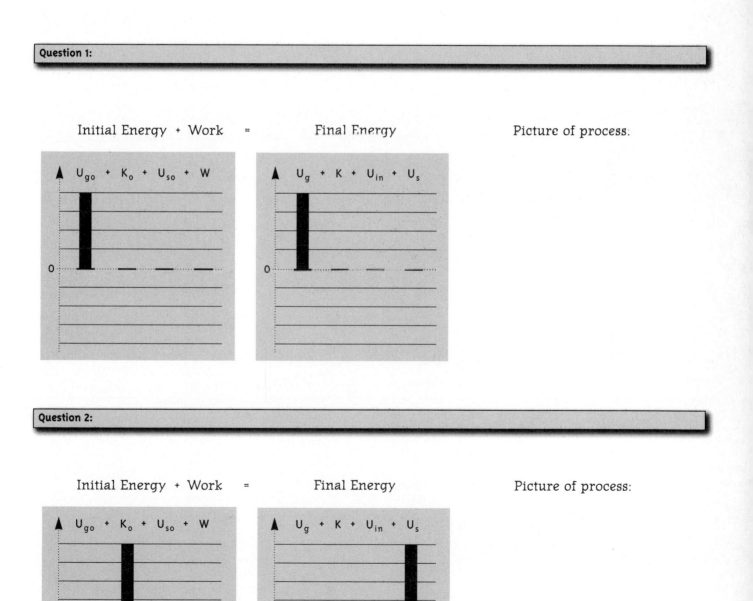

Initial Energy + Work = Final Energy   Picture of process:

U_{go} + K_o + U_{so} + W U_g + K + U_{in} + U_s

0   0

Question 2:

Initial Energy + Work = Final Energy   Picture of process:

U_{go} + K_o + U_{so} + W U_g + K + U_{in} + U_s

0   0

Question 3:

Initial Energy + Work = Final Energy   Picture of process:

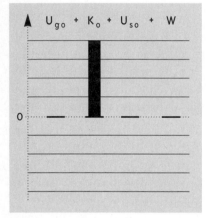

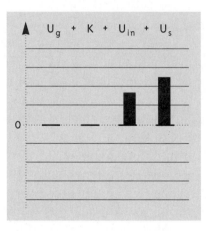
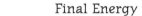

Question 4:

Initial Energy + Work = Final Energy   Picture of process:

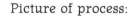

Question 5:

Initial Energy + Work   =   Final Energy   Picture of process.

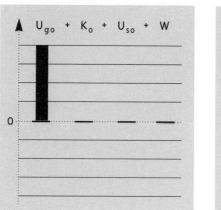

$$U_{go} + K_o + U_{so} + W$$

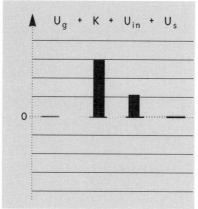

$$U_g + K + U_{in} + U_s$$

A 60-kg elevator initially moving up at 4.0 m/s slows to a stop in a distance of 4.0 m. Use the generalized work-energy equation to determine the tension in the elevator cable. Assume that the gravitational constant is 10 N/kg.

Question 1: Complete the work-energy bar charts.

Initial Energy + Work   =   Final Energy

$$U_{go} + K_o + U_s + W$$

$$U_g + K + U_s + U_{in}$$

0

0

Question 2: Apply the generalized work-energy equation

Solve for the cable tension.

A 60-kg elevator initially moving down at 4.0 m/s slows to a stop in a distance of 5.0 m. Use the generalized work-energy equation to determine the tension in the elevator cable. Assume that the gravitational constant is 10 N/kg.

Question 1: Complete the work-energy bar charts.

Initial Energy + Work   =   Final Energy

U_{go} + K_o + U_{so} + W

U_g + K + U_s + U_{in}

0

0

Question 2: Apply the generalized work-energy equation

Solve for the cable tension.

You are asked to help design an inverse bungee-jumping system. A 50-kg woman starts at rest with the top of her head 4.0 m below the support for the spring above. You are to choose the force constant for the spring so that after release she just makes it to but does not hit the support. The spring is initially stretched 3.5 m, and the gravitational constant is 10 N/kg.

Question 1: Complete the work-energy bar charts.

Initial Energy + Work = Final Energy

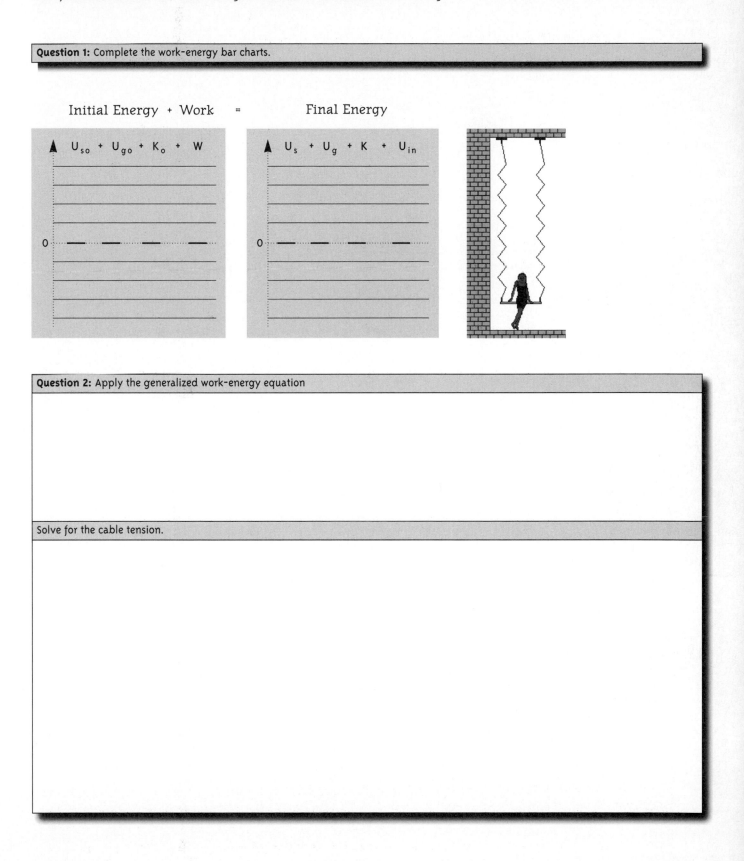

$U_{so} + U_{go} + K_o + W$

$U_s + U_g + K + U_{in}$

Question 2: Apply the generalized work-energy equation

Solve for the cable tension.

A 60-kg woman is sitting on a cushioned chair that is attached to a compressed spring and rests on a horizontal floor. The coefficient of kinetic friction between the chair and the floor is 0.10. When the spring is released, the woman is to slide 19.8 m to a glass of water. If she goes too far, she and the glass fall off the floor. If she stops too soon, she misses the glass. Determine the force constant for the spring that will cause her to stop at the glass. The spring is initially compressed 5.0 m, and the gravitational constant is 10N/kg.

Question 1: Complete the work-energy bar charts.

Initial Energy + Work   =   Final Energy

U_{so} + K_o + U_{go} + W

0

U_s + K + U_g + U_{in}

0

Question 2: Apply the generalized work-energy equation. | Solve for the magnitudes of the normal and friction forces.

Solve for the desired spring force constant.

Adjust the force constant of the spring to the predicted value and run the simulation to check your answer.

Estimate the effective force constant of the spring that launches the 16.5 g hot wheels car. The spring is initially streched to 15.5 cm. The car lands 6 m from the table. You will have to make estimates for any other quantities that are needed.

Concept(s) to be used:

Known or estimated quantities:

Unknown to be determined:

Calculations:

The 60-kg skier starts at rest on a hill inclined at an angle of 41.4° below the horizontal. The coefficient of kinetic friction between the skis and the snow is 0.20 (a sticky hill). Determine the speed after the skier travels 200 m down the hill. The gravitational constant is 10 N/kg. Start by determining the magnitudes of the normal and friction forces. Then construct initial and final qualitative energy bar charts for the motion. The origin of coordinates of the vertical axis is at the finish line.

Questions 1-2: Draw a force diagram and apply Newton's Second Law to determine the normal and friction forces.

Question 3: Complete the work-energy bar charts.

Initial Energy + Work = Final Energy

U_{go} + K_o + U_{so} + W

U_g + K + U_s + U_{in}

0

0

Apply the generalized work-energy equation and solve for the skier's speed.

The Hot Wheels car starts 1 m above the tabletop. Is its landing place on the floor consistent with the principles of physics?

Concept(s) to be used:

Known or estimated quantities:

Unknown to be determined:

Calculations:

A young man starts at rest at the top of a ride on a swing. Based on your observations and the application of the principles of physics, estimate the tension in each cable of the swing as it passes through the lowest point in it's swing.

Concept(s) to be used:

Known or estimated quantities:

Unknown to be determined:

Calculations:

A 50-kg skier starts at rest and is traveling at 1.5 m/s after moving 500 m up a hill inclined at an angle of 37° above the horizontal. The coefficient of kinetic friction between the skis and the snow is 0.20 (a pretty sticky hill). Determine the tension in the ski rope. The gravitational constant is 10 N/kg. Will the skier be able to hold onto the rope?

Questions 1-2: Draw a force diagram and apply Newton's Second Law to determine the normal and friction forces.

Question 3: Complete the work-energy bar charts.

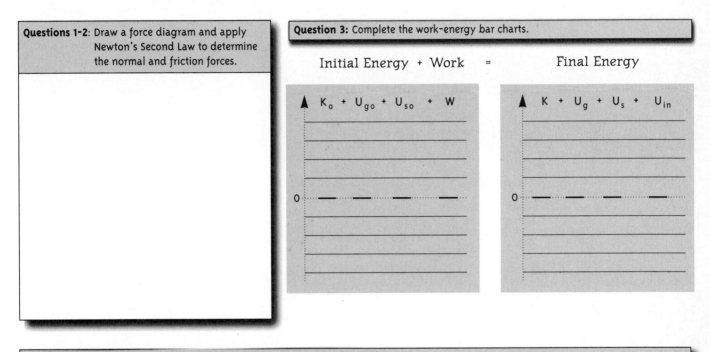

Initial Energy + Work = Final Energy

$$K_o + U_{go} + U_{so} + W$$

$$K + U_g + U_s + U_{in}$$

Apply the generalized work-energy equation and solve for the skier's speed.

Compare your answer to the value on the simulation.

A 200-kg ski cart starts against a compressed spring. When the spring is released, it launches the cart up the 31° hill so that the cart stops about 50 m from its starting position. The spring is initially compressed 10 m. Determine the force constant of the spring needed to achieve this goal. Try the problem first for a frictionless hill and then for a hill that has a coefficient of kinetic friction between the hill and the skis equal to 0.20. The gravitational constant is 10 N/kg.

Draw a force diagram and apply Newton's Second Law to determine the normal and friction forces.

Question 1 and Question 3: Complete the work-energy bar charts.

Initial Energy + Work = Final Energy

$$U_{so} + U_{go} + K_o + W$$

0

$$U_s + U_g + K + U_{in}$$

0

Apply the generalized work-energy equation and solve for the force constant of the spring.

Adjust the force constant to your answer and run the simulation.

With the sliders set as described for Question 2, a 10-kg yellow block hangs at one end of a cord. The 5.0-kg red block is attached to the other end and slides on the horizontal surface with a coefficient of kinetic friction of 0.20. The blocks start at rest. Determine their speed after the yellow block falls 4.0 m. The gravitational constant is 10 N/kg.

Question 2 and 4: Complete the work-energy bar charts.

Initial Energy + Work = Final Energy

$$K_{1o} + U_g + K_{2o} + W$$

$$K + U_{in} + U_g + K$$

0

0

Question 1: Determine the magnitude of the friction force.

Question 3 and 5: Apply the generalized work-energy equation.

Solve the equation.

You can follow a similar process to solve Questions 5-8.

Determine the speed of the 60-kg bricklayer after he rises 7.0 m. The 64-kg stack of bricks falls 7.0 m and is just ready to hit the ground. Assume that the gravitational constant is 10 N/kg. Ignore air resistance, friction in the pulley, and the pulley mass.

Question 1: Complete the work-energy bar charts.

Initial Energy + Work = Final Energy

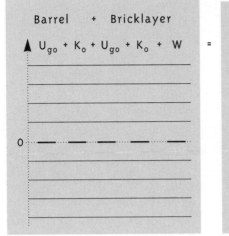

Barrel + Bricklayer

$U_{go} + K_o + U_{go} + K_o + W$ =

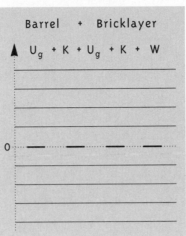

Barrel + Bricklayer

$U_g + K + U_g + K + W$

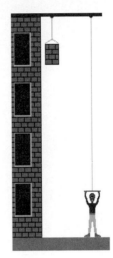

Question 2: Apply the generalized work-energy equation.

Solve the equation for the bricklayer's speed.

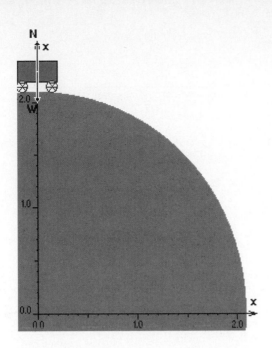

A 10-kg cart starts at rest at the top of a frictionless circular hill and coasts at increasing speed toward the right and down the hill. The radius of the circular path followed by the cart is 2.10 m. Determine the x- and y-coordinate positions when the cart leaves contact with the hill. The origin of the coordinates is at the center of the circle. Assume that the gravitational constant is 10 N/kg.

Question 1 — Solution Plan

• Describe the parts of the problem.
• Identify the best physics principle to apply for each part.
• Do you need any other concepts?

5.14 Cart Leaves Circular Hill continued

Question 2: Complete the work-energy bar charts.

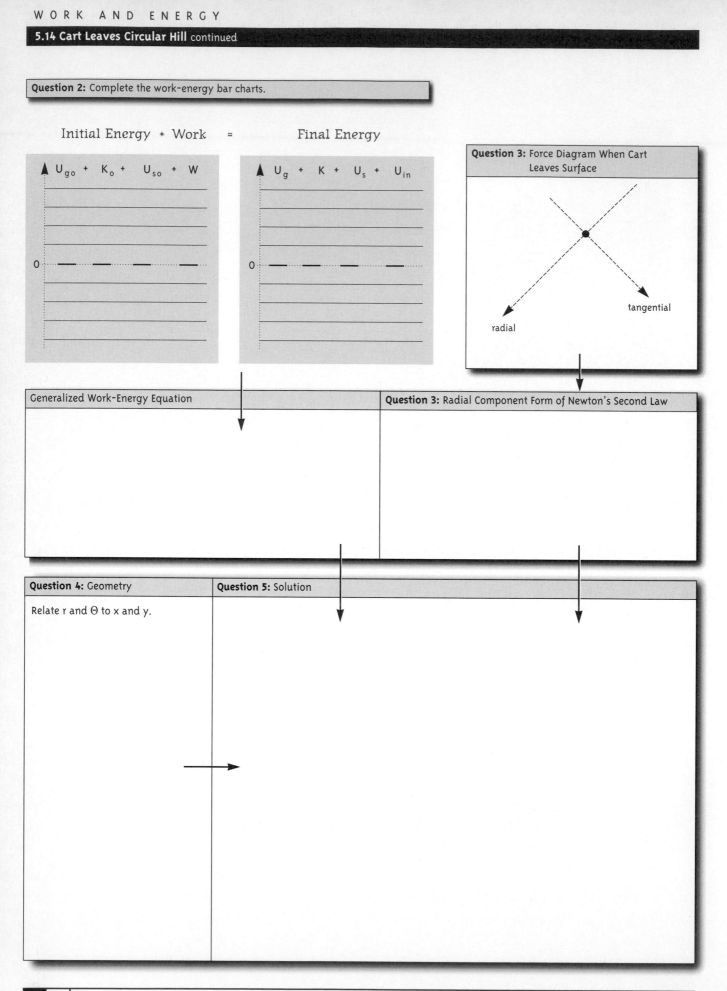

Initial Energy + Work = Final Energy

$U_{go} + K_o + U_{so} + W$

$U_g + K + U_s + U_{in}$

0

0

Question 3: Force Diagram When Cart Leaves Surface

radial

tangential

Generalized Work-Energy Equation

Question 3: Radial Component Form of Newton's Second Law

Question 4: Geometry

Relate r and Θ to x and y.

Question 5: Solution

6

MOMENTUM

Finish
Line

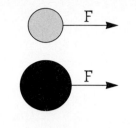

Question 1: Two pucks initially at rest are pushed by identical 10-N forces caused by an object not shown. The blue puck has mass 1.0 kg, and the orange puck has mass 2.0 kg. Which puck reaches the finish line first?

Blue Puck   Orange Puck   It's a tie

Question 2: The pucks start again and travel until the blue puck reaches the finish line. At that time, which puck will have the greater change in momentum? Explain your choice.

Blue Puck   Orange Puck   They have equal changes.

Question 3: The pucks start again and travel until the blue puck reaches the finish line. At that time, which puck will have the greater change in kinetic energy? Explain your choice.

Blue Puck   Orange Puck   They have equal changes.

Question 4: Record the kinetic energy change when the blue puck reaches the finish line. When the green puck reaches the finish line, will it have more, the same, or less kinetic energy change than the blue puck had when it reached the finish line? Explain your choice.

dK greater for Orange Puck   dK same for both pucks   dK greater for Blue Puck

Question 5: Record the momentum change when the blue puck reaches the finish line. When the green puck reaches the finish line, will it have more, the same, or less momentum change than the blue puck had when it reached the finish line? Explain your choice.

dp greater for Orange Puck   dp same for both pucks   dp greater for Blue Puck

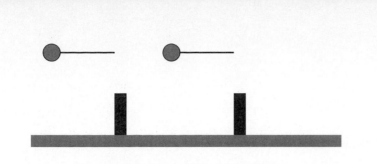

Question 1: Two identical balls swing down and hit equal mass bricks. The collision on the left is totally elastic and that on the right is totally inelastic. First observe the behavior of the balls when the bricks are anchored so that they cannot move.

Question 2: When the anchors holding the bricks are removed and the balls swing down to hit the bricks, which brick is most likely to be knocked over? Explain your choice.

Elastic collision   Inelastic collision   Equal chance

Question 3: Which collision causes the greater impulse on the brick? Explain your choice.

Elastic collision   Inelastic collision   Equal chance

Question 4: Which collision causes the greater change in momentum of the ball? Remember that momentum is a vector quantity.

Elastic collision   Inelastic collision   Equal chance

Are the answers to Questions 3 and 4 related in any way?

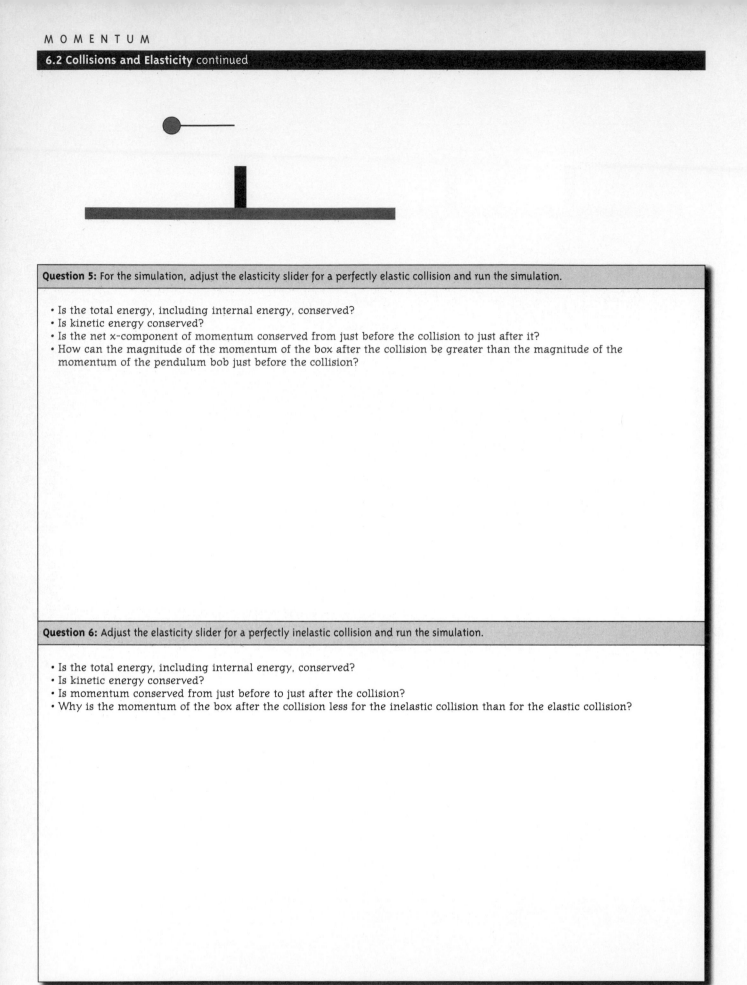

Question 5: For the simulation, adjust the elasticity slider for a perfectly elastic collision and run the simulation.

• Is the total energy, including internal energy, conserved?
• Is kinetic energy conserved?
• Is the net x-component of momentum conserved from just before the collision to just after it?
• How can the magnitude of the momentum of the box after the collision be greater than the magnitude of the momentum of the pendulum bob just before the collision?

Question 6: Adjust the elasticity slider for a perfectly inelastic collision and run the simulation.

• Is the total energy, including internal energy, conserved?
• Is kinetic energy conserved?
• Is momentum conserved from just before to just after the collision?
• Why is the momentum of the box after the collision less for the inelastic collision than for the elastic collision?

For the first simulation, adjust each slider, one at a time, to get a feel for its effect on the collision of the orange puck with the blue puck. Shortly, you will examine the readings on one or more meters and invent rules that apply to these collisions. The pucks slide on a frictionless surface.

Questions 1-2 — The x-Component of Momentum: This simulation indicates the x-component of momentum ($m\,v_x$) of each puck. Run the simulation several times and adjust the elasticity, mass ratio, and y-position before each run. Based on your observations, invent a rule that seems to describe for the x-component of momentum during these collisions.

Question 3 — The y-Component of Momentum: The next simulation indicates the y-component of momentum ($m\,v_y$) of each puck. Run the simulation several times and adjust the y-position, the elasticity, and the mass ratio before each run. Based on your observations for a variety of conditions, invent a rule that seems to describe the variation of the y-component of momentum during the collisions.

Questions 4-5 — Elasticity and Kinetic Energy During Collisions: The next simulation has energy bar charts that indicate by the lengths of bars the initial and final kinetic energy and internal energy of the two-puck system. Run the collisions under a variety of conditions by adjusting the sliders one at a time before each run. Then indicate

(a) under what conditions the total energy, including internal energy, is conserved

(b) under what condition(s) the total kinetic energy (the sum of the kinetic energy of the two pucks) is conserved during the collisions.

These collisions with conserved kinetic energy are called **elastic** collisions. If kinetic energy is not conserved, the collision is said to be partially or totally **inelastic.**

(a) Conditions for energy conservation (including internal energy) during collisions:

(b) Conditions for kinetic energy conservation during collisions:

SUMMARY

• If the net force in the x-direction exerted by other objects on a system of two or more colliding objects is zero, then the sum of the x-components of the momentum of the objects in the system is conserved or constant.

• If the net force in the y-direction exerted by other objects on a system of two or more colliding objects is zero, then the sum of the y-components of the momentum of the objects in the system is conserved or constant.

• If no external forces do work on the objects in the system (like the two pucks), then the total energy of the system is conserved. There may be considerable conversion of kinetic energy to internal energy.

• For so-called elastic collisions, the kinetic energy of the system is conserved.

Show that both energy and momentum cannot be conserved if the number of balls that swing down differs from the number that swing up after the collision. Now aren't they smart.

Concept(s) to be used:

Known or estimated quantities:

Unknown to be determined:

Calculations:

Question 1: A 1.0-kg green puck travels at 4.0 m/s in the positive x-direction and has a head-on collision with a 0.5-kg blue puck, initially at rest. The collision is totally inelastic — they stick together. Predict the velocity of the two pucks after the collision. How much internal energy is produced during the collision?

Question 2: A 1.0-kg green puck travels at 4.0 m/s in the positive x-direction and has a head-on totally elastic collision with a 0.5-kg blue puck, initially at rest. Predict the velocity of each puck after the collision. How much internal energy is produced during the collision?

Question 3: A 1.0-kg green puck traveling in the positive x-direction at 4.0 m/s has a glancing collision with a 1.0-kg blue puck. The collision is elastic. With the x and y-velocity components for the blue puck left off, run the simulation and observe the x- and y-velocity components of the green puck. Then predict the x- and y-components of velocity of the blue puck. After your calculations, click on the blue puck's velocity components and rerun the simulation to check your answers. Is kinetic energy conserved? Support your answer.

Question 1: A 2000-kg blue truck initially travels in the positive x-direction at 10 m/s, and a 1000-kg yellow car initially travels in the positive y-direction at 20 m/s. The collision is totally inelastic (elasticity 0.0). Determine all the final velocity components for the two vehicles. How much internal energy is produced during the collision?

Question 2: A 2000-kg blue truck travels in the positive x-direction at 10 m/s, and a 1000-kg yellow car travels in the positive y-direction at 20 m/s. The collision is totally elastic. Turn on the x-component velocity for the truck and the y-component velocity for the car. Run the simulation and use momentum conservation principles to determine all the final velocity components for the vehicles. How much internal energy is produced during the collision? Is this possible for real car collisions?

Question 3: A 2000-kg blue truck travels in the positive x-direction at 24 m/s, and a 1000-kg yellow car initially travels in the positive y-direction at 12m/s. They have a totally inelastic collision. Turn off all the velocity components. Determine all of the final velocity components for the two vehicles. How much internal energy is produced during the collision?

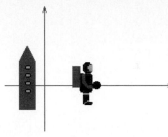

Question 1: A 60-kg astronaut drifts away from a spaceship at speed 2.0 m/s. She has no jets to help her return. When she is 20 m from the ship, she flings her 30-kg auxiliary oxygen supply into space. The oxygen supply now moves at 10 m/s relative to the spaceship. Does she have a reasonable chance of returning to the ship alive?

Determine her velocity just after flinging the oxygen supply.

Will she make it back to the ship in a reasonable time? Justify your answer.

Question 1: A 3.0-kg projectile traveling horizontally through the air at 3.0 m/s explodes into a 2.0-kg blue part and a 1.0-kg orange part. Each part experiences a 3.0 kg m/s impulse (positive on the blue part and negative on the orange). After the explosion, they take 1.0 s to reach the ground. If they are at position x = 0.0 m at the time of the explosion, where will the two parts land?

Question 2: A 3.0-kg projectile traveling horizontally through the air at 3.0 m/s explodes into a 1.0-kg blue part and a 2.0-kg orange part. Each part experiences a 3.0 kg m/s impulse (positive on the blue part and negative on the orange). After the explosion, they take 1.0 s to reach the ground. If they are at position x = 0.0 m at the time of the explosion, where will the two parts land?

Question 3: A 1.0-kg projectile traveling horizontally through the air at 3.0 m/s explodes into a 0.5-kg blue part and a 0.5-kg orange part. Each part experiences a 3.0 kg m/s impulse (positive on the blue part and negative on the orange). After the explosion, they take 1.0 s to reach the ground. If they are at position x = 0.0 m at the time of the explosion, where will the two parts land?

Question 4: A 1.0-kg projectile traveling horizontally through the air at 3.0 m/s explodes into a 0.5-kg blue part and a 0.5-kg orange part. Each part experiences a 2.0 kg m/s impulse (positive on the blue part and negative on the orange). After the explosion, they take 1.0 s to reach the ground. If they are at position x = 0.0 m at the time of the explosion, where will the two parts land?

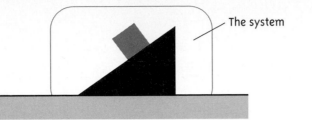

— The system

Answer the following questions concerning the block and incline. All surfaces have negligible friction.
Assume that the gravitational constant is 10 N/kg = 10 m/s².

Question 1 — The System: Consider the block and incline as the system. Are there any external forces exerted on the system that have non-zero x-components? If so, describe these forces.

Question 2 — Mass of Incline: Run the simulation with the x-components of velocity turned on. Based on the meter readings after the 10-kg block leaves the incline and using the conservation of momentum principle, determine the mass of the incline.

Question 3 — Another Momentum Problem: Reset the simulation and change the block's mass to 2.00 kg. Turn off the block's x-component of velocity and run the simulation. Based on your observations, predict the x-component of velocity of the block after it leaves contact with the incline. Run the simulation again with block x on to check your work.

Question 4 — Force Diagram for Block and Incline: Construct a force diagram for the block and incline together during the time interval that the block and incline are moving and the block is about halfway down the incline. Use vertical and horizontal axes.

y

x

6.8 Block Sliding Down an Incline continued

Question 4 — Apply Newton's Second Law: Apply the y-component form of Newton's Second Law ($\Sigma F_y = ma_y$) to the force diagram you drew. Show that the meter readings for the forces in the simulation are consistent with this law. Assume that the gravitational constant is 10 N/kg = 10 m/s^2.

Question 5 — Force Diagram for Block: Construct a force diagram for the block when it is moving, and it is about halfway down the incline. Use vertical and horizontal axes. Then apply the x- and y-component forms of Newton's Second Law to the force diagram. Use the meter readings and the Second Law equations to determine the angle of the incline. Each equation should produce the same answer. Assume that the gravitational constant is 10 N/kg = 10 m/s^2.

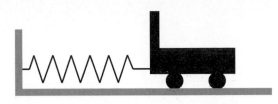

After coming down a ski slope, a 50-kg skier traveling at 12 m/s, runs into a 25-kg padded cart. Velcro fasteners cause the skier and cart to join together. After joining together, the cart and skier compress a 400-N/m spring that is initially relaxed. The skier gets a nice vibro-ride at the end of the ski run. Your task is to determine the maximum distance that the spring will compress before stopping the cart and skier. The surface is frictionless.

Before starting your work, construct qualitative energy bar charts for the following times:

Just before collision	Just after collision before compression	At maximum compression of spring
K_{so} + K_{co} + U_{so} + U_{in}	K_s + K_c + U_s + U_{in}	K_s + K_c + U_s + U_{in}

Concepts for the Parts: Break the big problem into parts, indicate the unknown to be determined for each part, and identify the concept that will be most useful for analyzing each part.

Question 1: Part I — The Collision: You might be tempted to use energy conservation for the whole problem. This does not work well for the collision because of the production of an unknown amount of internal energy during the collision — the skier might even break a bone or get some black and blue marks if the padding is not sufficient. However, if there are no external forces in the horizontal direction on the skier-cart system, then the x-component of momentum is conserved. Use that principle to determine the speed of the cart and skier together at the instant just after the collision.

Question 2: Part II — Compressing the Spring: You found that the cart and skier's speed together at the instant after the collision was 8.0 m/s. Now use another important conservation principle to determine the maximum distance that the 400-N/m spring gets compressed before stopping the skier and cart.

Question 3: After the collision, at what position does the skier feel the greatest force from the padded cart?
(a) just after the collision when moving fastest
(b) at the instant the cart stops when the spring is compressed the most
(c) when the spring is about half compressed
Justify your answer.

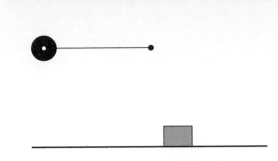

A pendulum bob swings down and hits a box. The box slides on a surface and comes to a stop at some unknown distance from its starting point. You are to determine that unknown distance given the following information:

- mass of pendulum bob is 0.50 kg
- mass of box is 1.0 kg
- length of pendulum string is 3.0 m
- coefficient of kinetic friction between the box and surface is 0.10

Construct qualitative work-energy bar charts for the following times.
The system includes the pendulum, the box, the surface, and the mass of the earth.

At start of process

Pend. Bob   Box

$K_o + U_{go} +$ K_o

0

Just before the collision

Pend. Bob   Box

$K + U_g$ + $K + U_{int}$

0

Just after the collision

Pend. Bob   Box

$K + U_g$ + $K + U_{int}$

0

Question 1: Determine the speed of the pendulum bob just before it hits the box.

When box has stopped at end

Pend. Bob   Box

$K + U_g$ + $K + U_{int}$

0

Question 2: Determine the speed of the box immediately after the pendulum bob hits it. (Why is energy conservation not appropriate for this calculation?)

Question 3: Determine the distance that the box travels after the collision.

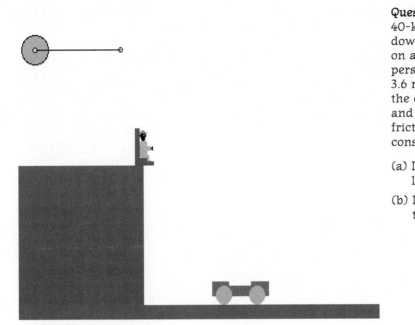

Question 1: A 3.0-m-long rope with a 40-kg medicine ball on the end swings down and hits an 80-kg person sitting on a ledge. The ball stops, and the person flies off and lands in a cart 3.6 m below (if the cart is located at the correct position). The person and cart then slide happily across a frictionless surface. The gravitational constant is 10 N/kg.

(a) Decide where the cart should be located so that the person lands in it.

(b) Decide how fast the cart moves after the person has settled into it.

RECOMMENDED STRATEGY:

• Identify the parts of the problem.
• Identify the unknown to be determined for each part.
• Identify the main principle to be used to determine that unknown for each part.
• Solve the parts and reassemble to answer the big problem.

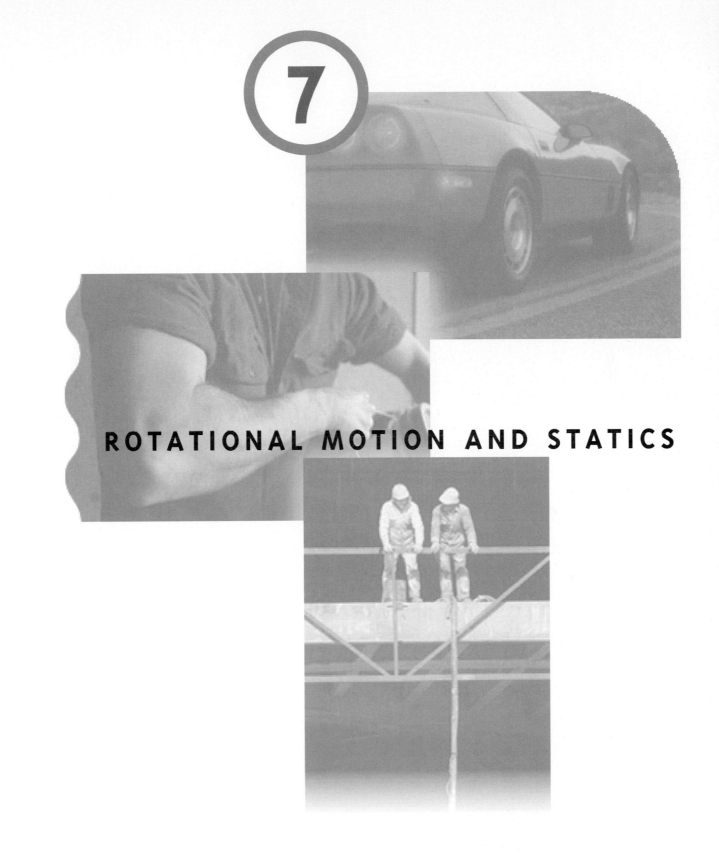

7

ROTATIONAL MOTION AND STATICS

A rope slanted at 37° above the horizontal supports the right end of a 10-kg, 4.0-m-long horizontal beam. A pin through the beam supports its left end. If a 20-kg brick sits on the beam 3.0 m from the left end, the rope tension is 333.3 N. The gravitational constant is 10 N/kg.

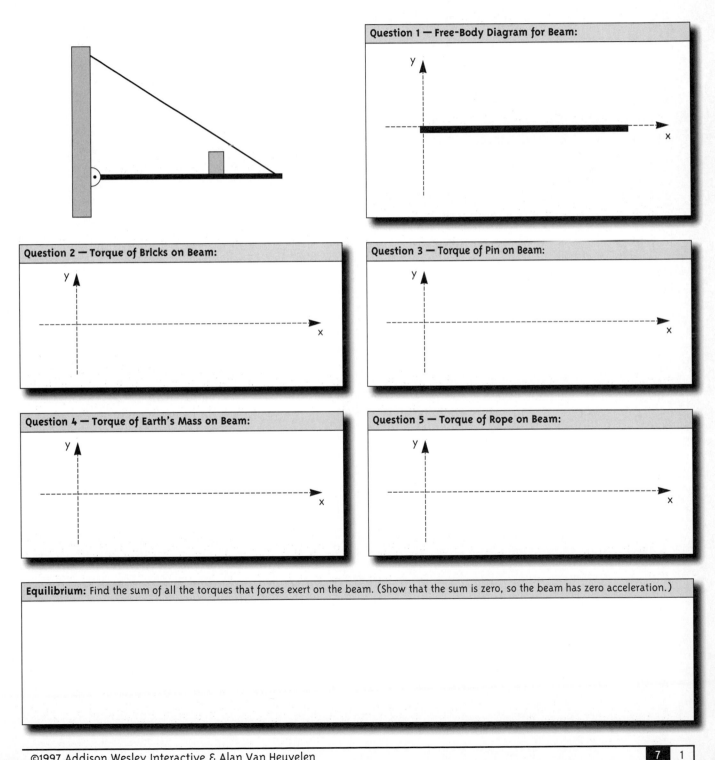

Question 1 — Free-Body Diagram for Beam:

y

x

Question 2 — Torque of Bricks on Beam:

y

x

Question 3 — Torque of Pin on Beam:

y

x

Question 4 — Torque of Earth's Mass on Beam:

y

x

Question 5 — Torque of Rope on Beam:

y

x

Equilibrium: Find the sum of all the torques that forces exert on the beam. (Show that the sum is zero, so the beam has zero acceleration.)

A horizontal rope with tension 237 N supports the right end of a 4.0-m-long beam that is slanted 30° above the horizontal. A pin through the beam supports its left end. A 10-kg brick hangs from the beam 3.46 m from the pin supporting the beam's left side. The gravitational constant is 10 N/kg.

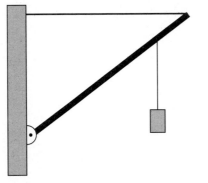

Free-Body Diagram for Beam:

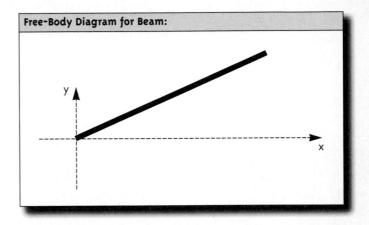

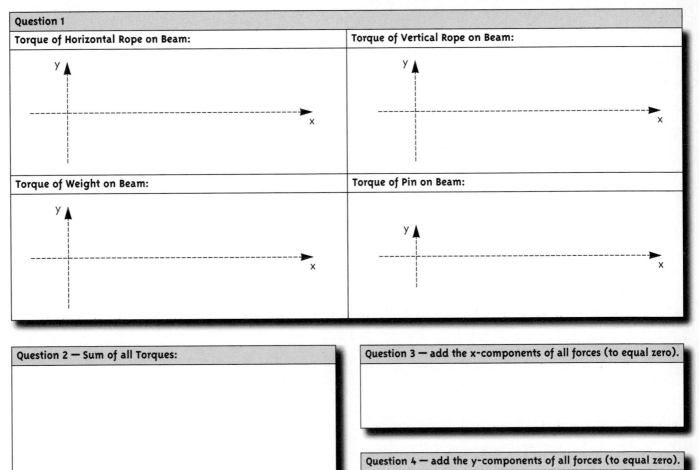

Question 1

Torque of Horizontal Rope on Beam:

Torque of Vertical Rope on Beam:

Torque of Weight on Beam:

Torque of Pin on Beam:

Question 2 — Sum of all Torques:

Question 3 — add the x-components of all forces (to equal zero).

Question 4 — add the y-components of all forces (to equal zero).

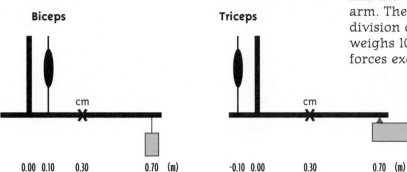

Biceps   **Triceps**

The simulations in the activity provide models for how the biceps and triceps work. The horizontal beam represents the lower arm, and the vertical beam represents the upper arm. The lower arm has a mass of 10 kg. Each division on the screen is 0.20 m. The block weighs 10 kg. Determine the muscle tension and forces exerted on the lower arm.

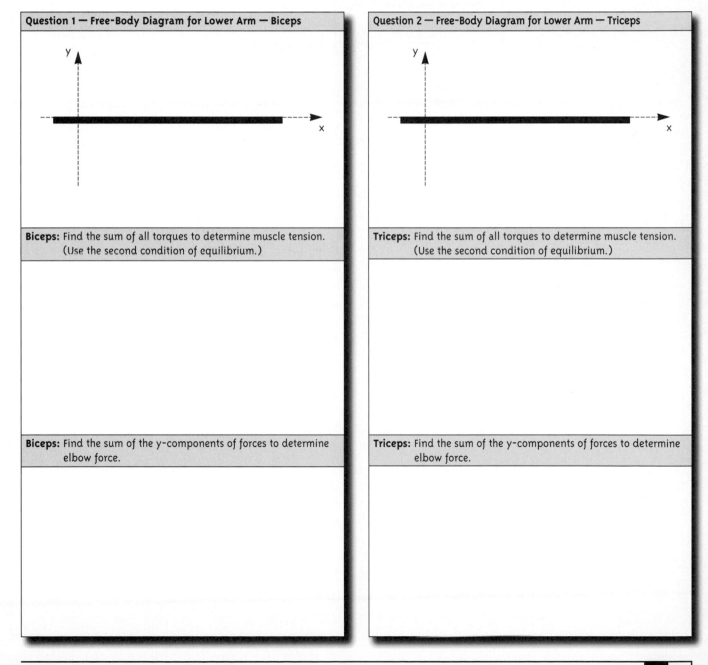

Question 1 — Free-Body Diagram for Lower Arm — Biceps

Biceps: Find the sum of all torques to determine muscle tension. (Use the second condition of equilibrium.)

Biceps: Find the sum of the y-components of forces to determine elbow force.

Question 2 — Free-Body Diagram for Lower Arm — Triceps

Triceps: Find the sum of all torques to determine muscle tension. (Use the second condition of equilibrium.)

Triceps: Find the sum of the y-components of forces to determine elbow force.

Estimate the mass of the meter stick that has been cut in half. The fulcrum is 1 cm from one end of the stick, the "biceps" support cable is 9 cm from the fulcrum, and the 100-g, 200-g, and 300-g weight hang 48 cm from the fulcrum.

Concept(s) to be used:

Known or estimated quantities:

Unknown to be determined:

Calculations:

Two painters stand on a 10-kg, 4.0-m-long uniform beam that is supported by ropes on each end. The gravitational constant is 10 N/kg. Determine the tension in each rope. (Complete the information below to answer Question 1.)

Description of Situation:

• The mass of the painter on the left is _____ kg, and he stands _____ m from the left rope.

• The mass of the painter on the right is _____ kg, and she stands _____ m from the right rope.

Free-Body Diagram for Beam (draw axes):

First Condition of Equilibrium (y-components):

Second Condition of Equilibrium:

Complete Solution:

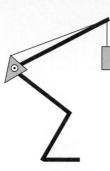

A 1.00-m-long beam represents the backbone. A 20-kg load hangs from the end. The beam has a 30-kg uniform mass. The back muscle connects to the beam 0.77 m from the pivot point at the lower left of the beam. The back muscle makes a 9.75° angle with respect to the beam. The gravitational constant is 10 N/kg.

Free-Body Diagram for Backbone (Beam)

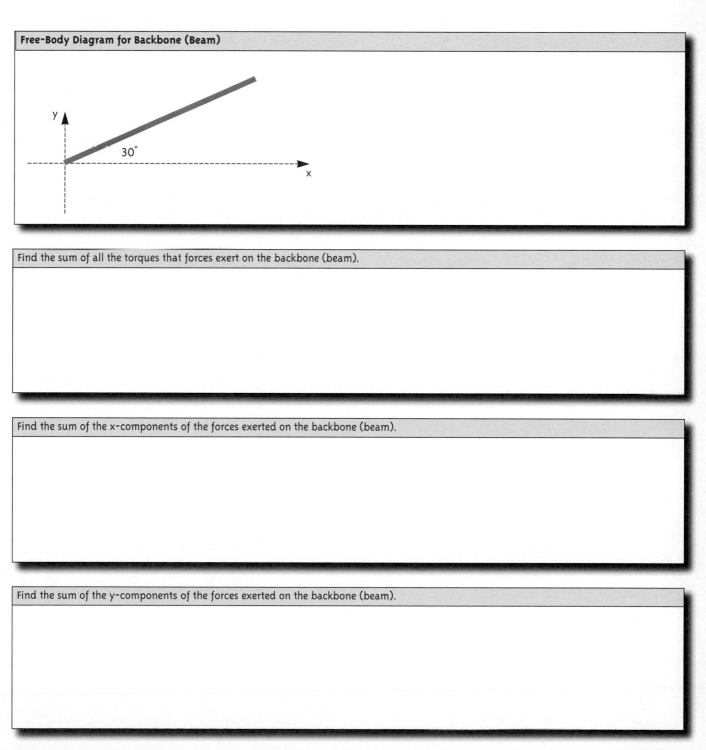

Find the sum of all the torques that forces exert on the backbone (beam).

Find the sum of the x-components of the forces exerted on the backbone (beam).

Find the sum of the y-components of the forces exerted on the backbone (beam).

Estimate the tension in the arm of the 170-lb person swinging the bucket of water. Stop the movie about two-thirds of the way through when the person is facing you and use statics for your estimate.

Concept(s) to be used:

Known or estimated quantities:

Unknown to be determined:

Calculations:

A professor sits 2.0 m from the left fulcrum of a 5.0-m-long, 20-kg uniform beam. A rope connected to the other end of the beam passes up over a pulley and down to a harness worn by the professor. The rope makes a 45° angle with the beam. The gravitational constant is 10 N/kg. Determine the rope tension and the normal force of the beam on the professor. (Complete the following information to answer Question 1.)

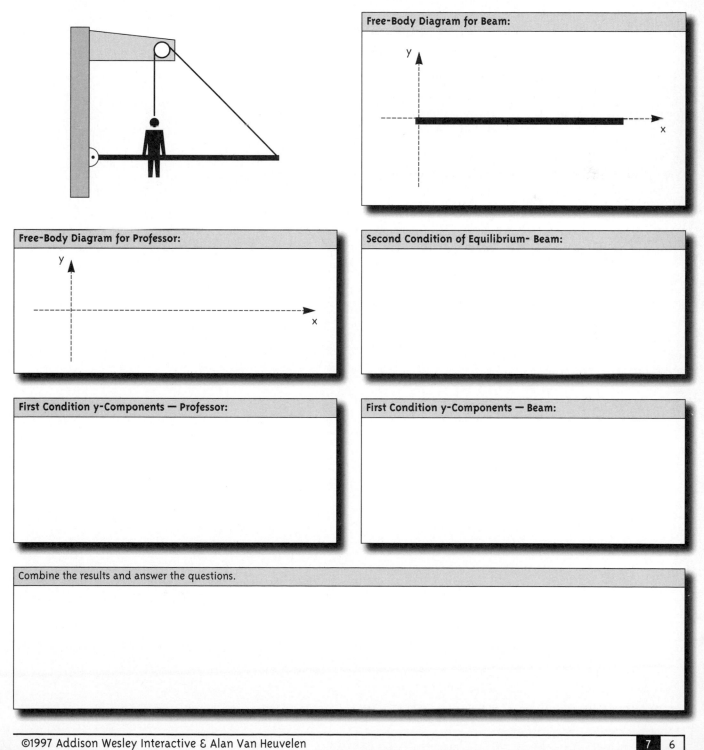

Free-Body Diagram for Beam:

y

x

Free-Body Diagram for Professor:

y

x

Second Condition of Equilibrium- Beam:

First Condition y-Components — Professor:

First Condition y-Components — Beam:

Combine the results and answer the questions.

Determine the rotational inertia of each beam of four balls shown below. The balls rotate about an axis perpendicular to and through
(a) the right end of the beam and
(b) the center of the beam. Each ball has a mass of 1.0 kg, and the beam connecting the balls has no mass.

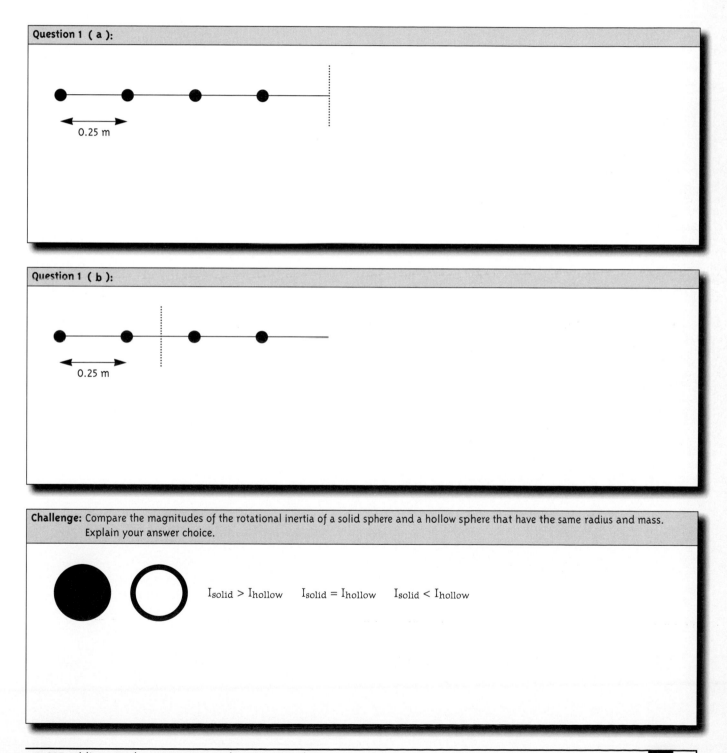

Question 1 (a):

0.25 m

Question 1 (b):

0.25 m

Challenge: Compare the magnitudes of the rotational inertia of a solid sphere and a hollow sphere that have the same radius and mass. Explain your answer choice.

$I_{solid} > I_{hollow}$ $I_{solid} = I_{hollow}$ $I_{solid} < I_{hollow}$

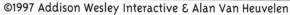

Each disk below rotates about an axis through the center of the disk. The word *faster* means that the disk is turning faster and faster and does not imply anything about how fast it is turning. Similarly, the word *slower* means that the disc is turning slower and slower (its angular speed is decreasing). Add vectors to show the direction of the angular velocity ω and the angular acceleration α for the different types of motion.

Angular Velocity ω:

Direction: _____
Sign: _____

Direction: _____
Sign: _____

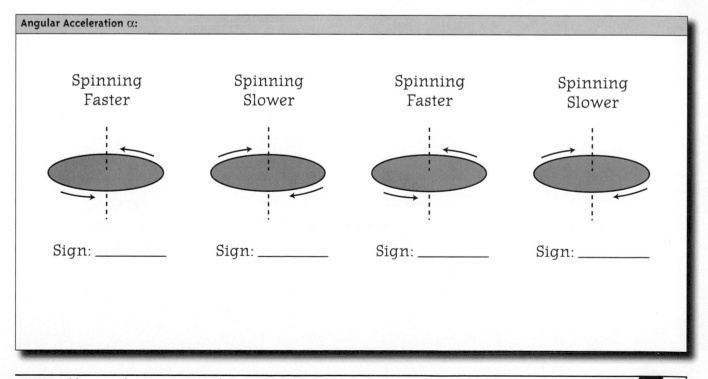

Angular Acceleration α:

Spinning Faster   Spinning Slower   Spinning Faster   Spinning Slower

Sign: _____   Sign: _____   Sign: _____   Sign: _____

A top view of the six disks from Question 1 is shown below. Each disk rotates about an axis through its center and perpendicular to the page. You are to run the simulation and determine the sign of the angular velocity ω and of the angular acceleration α for each disk. The choices are +, −, or 0. Note that for angular velocity ω:

- counterclockwise is positive (+)
- clockwise is negative (−)
- zero (0)

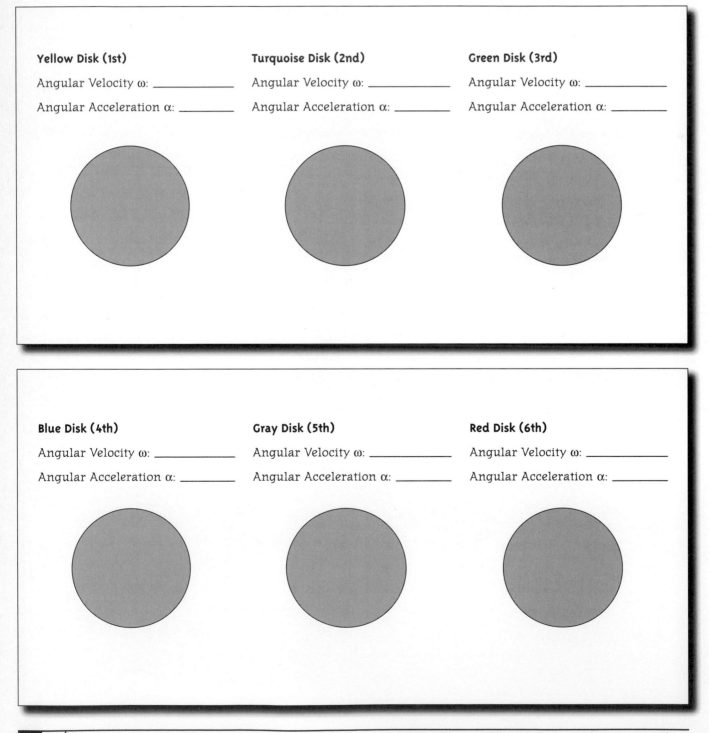

Yellow Disk (1st)

Angular Velocity ω: _____

Angular Acceleration α: _____

Turquoise Disk (2nd)

Angular Velocity ω: _____

Angular Acceleration α: _____

Green Disk (3rd)

Angular Velocity ω: _____

Angular Acceleration α: _____

Blue Disk (4th)

Angular Velocity ω: _____

Angular Acceleration α: _____

Gray Disk (5th)

Angular Velocity ω: _____

Angular Acceleration α: _____

Red Disk (6th)

Angular Velocity ω: _____

Angular Acceleration α: _____

The simulation shows the side view of a front-wheel-drive car. The right wheel on the simulation represents the front wheels, and the left wheel on the simulation represents the back wheels.

Question 1: The motor exerts a –400 N•m clockwise torque on the front wheels of the 100-kg car. Indicate the direction of the friction force of the road on the front wheels and the direction that the car accelerates.

Question 2: The motor exerts a +400 N•m counterclockwise torque on the front wheels of the 100-kg car. Indicate the direction of the friction force of the road on the front wheels and the direction that the car accelerates.

Question 3: When the car was accelerating backward (toward the left on the simulation), the normal force of the road was greater on the front wheels than on the back wheels. Explain. Think about Newton's Third Law and the torque of the motor on the wheel and the torque of the wheel on the motor (and car).

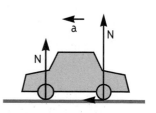

You are asked to determine the magnitude of the force (represented by the arrow) that a parent must apply on one pod of the rotoride so that the pod's angular acceleration is 0.40 rad/s^2. The 80-kg pod on each end, which includes the mass of the passenger, is connected by a uniform rod of rotational inertia 8.13 kg•m^2. This does not include the rotational inertia of the pods, which are located at each end 1.0 m from the center of the beam.

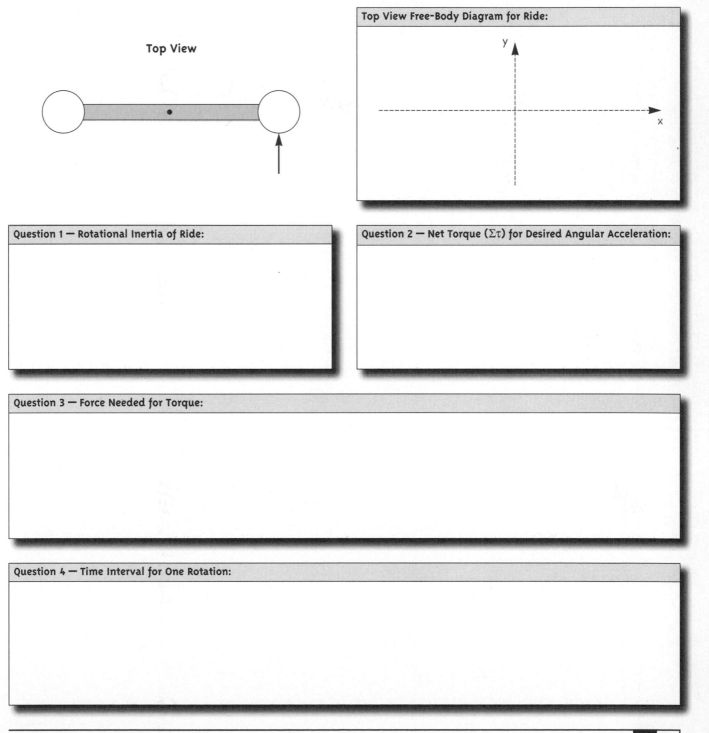

Top View

Top View Free-Body Diagram for Ride:

Question 1 — Rotational Inertia of Ride:

Question 2 — Net Torque ($\Sigma\tau$) for Desired Angular Acceleration:

Question 3 — Force Needed for Torque:

Question 4 — Time Interval for One Rotation:

A 5.0-kg, 4.0-m-long ladder falls. Determine the angular acceleration of the ladder at the instant it is tilted 1.0 radian above the horizontal. The ladder has a uniform mass distribution, and the gravitational constant is 10.0 N/kg.

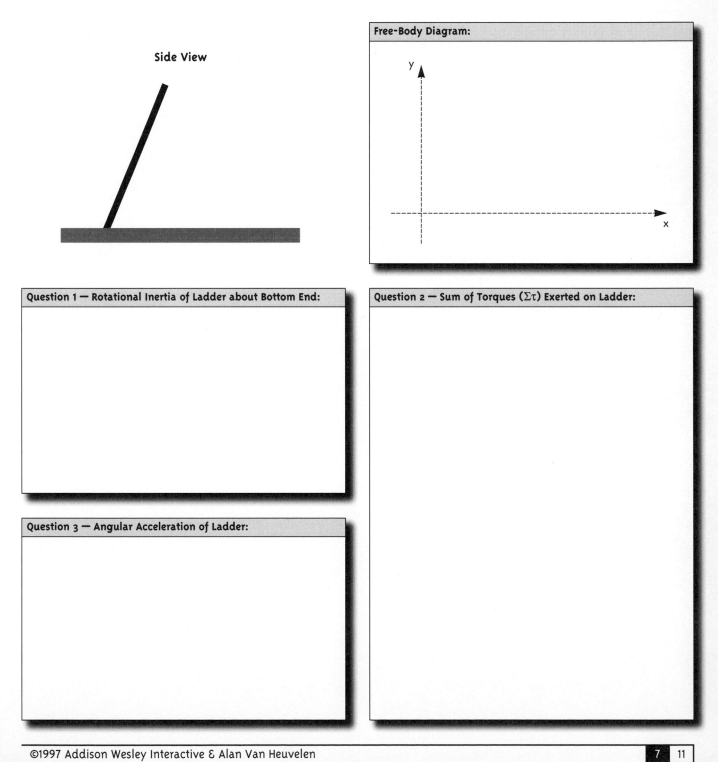

Side View

Free-Body Diagram:

y

x

Question 1 — Rotational Inertia of Ladder about Bottom End:

Question 2 — Sum of Torques (Στ) Exerted on Ladder:

Question 3 — Angular Acceleration of Ladder:

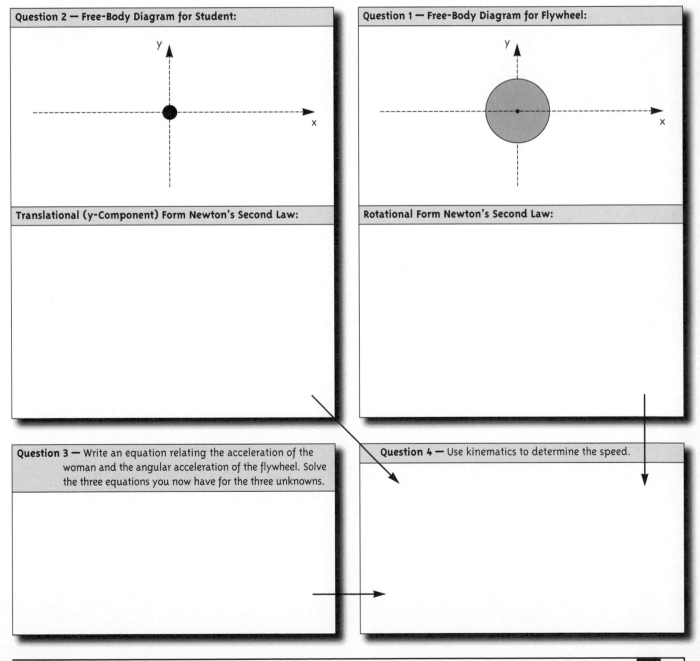

A 60-kg student hangs from a trapeze held by a rope wrapped around a 1.0-m-radius flywheel with a rotational inertia of 180 kg•m². She starts at rest. Determine her speed after falling 5.0 m towards the ground. The gravitational constant is 10 N/kg.

Question 2 — Free-Body Diagram for Student:

Translational (y-Component) Form Newton's Second Law:

Question 1 — Free-Body Diagram for Flywheel:

Rotational Form Newton's Second Law:

Question 3 — Write an equation relating the acceleration of the woman and the angular acceleration of the flywheel. Solve the three equations you now have for the three unknowns.

Question 4 — Use kinematics to determine the speed.

A solid block slides down a frictionless incline. A solid disk of the same mass rolls down an identical incline. Your goal is to decide which object first reaches the bottom of the incline. Before making your prediction, complete the work-energy bar charts for the block system and the disk system, starting with the object at the top of the incline and ending with it at the bottom. The kinetic energy can be in the form of translational kinetic energy (Kt) and/or rotational kinetic energy (Kr). Finally, make your choice and provide a reason for that choice. (Complete the information below to answer Questions 1 and 2.)

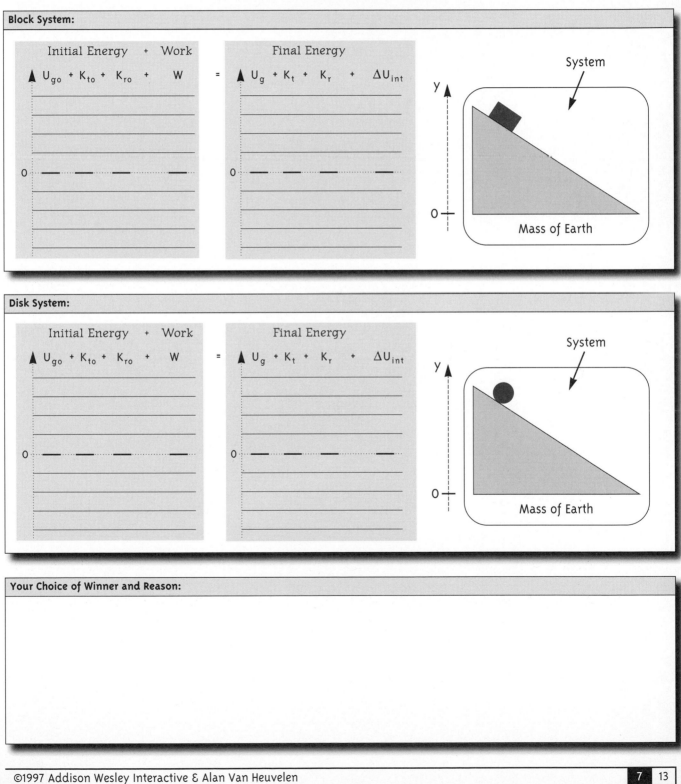

Block System:

Initial Energy + Work

$U_{go} + K_{to} + K_{ro} + W = U_g + K_t + K_r + \Delta U_{int}$

Final Energy

System

Mass of Earth

Disk System:

Initial Energy + Work

$U_{go} + K_{to} + K_{ro} + W = U_g + K_t + K_r + \Delta U_{int}$

Final Energy

System

Mass of Earth

Your Choice of Winner and Reason:

A 60-kg student hangs on a trapeze held by a rope wrapped around a 1.0-m-radius flywheel with a rotational inertia of 180 kg•m². The student starts at rest. Determine her speed after falling 5.0 m towards the ground. Assume that the gravitational constant is 10 N/kg.

Question 1 — Work-Energy Bar Chart:

Initial Energy   +   Work

U_{go} + K_{to} + K_{ro} +   W   = U_g + K_t + K_r + U_g

Final Energy

0

0

← System

Mass of Earth

Question 2 — Apply the energy conservation principle and the generalized work-energy equation.

Question 3 — Relate the translational speed of the student and the angular (rotational) speed of the flywheel.

Question 4 — Solution.

A rotoride has a uniform rod with 80-kg child-occupied pods on each end at 1.0 m from the center of the rod. A parent's continuous 67-N push on one pod causes the pods and rod to rotate at increasing angular speed. The rotational inertia of the rod-and-two-pod system is 168.1 kg·m². Use an energy approach to determine the time interval needed for the first complete turn of the rod, starting from rest. (Ignore friction.)

Question 1 — Work-Energy Bar Chart:

Initial Energy + Work

$$K_o + U_{go} + U_{so} + W = K + U_g + U_s + U_{in}$$

Final Energy

0

0

System includes the rotoride but not the pusher.

F

Question 2 — Apply the energy conservation principle and the generalized work-energy equation.

Question 3 — Determine the work done by the person pushing.

Question 4 — Find the final angular velocity.

Question 5 — Determine the time interval for one rotation.

Observe the behavior of the ball and bat in Question 1. Then complete the following information.

Question 2 — Angular Momentum of the Ball: Based on your observations, determine an expression for the angular momentum of the ball.

Question 3 — Angular Momentum of Extended Body: Based on your observations, determine an expression for the angular momentum of the bat.

Question 4 — Angular Momentum Conservation: Based on your observations, write a rule that describes the angular momentum of the ball-bat system during their collisions.

For each of the following situations, determine the missing quantity (or quantities) for the collision. The ball's mass is 0.50 kg, and the bat's rotational inertia is 3.0 kg·m².

Question Number:	5	6	7	Workspace:
Elasticity	1	1	0	
Initial x-velocity (m/s) ball	−4.0	−4.0	−4.0	
Initial y-position (m) ball	−3.0	−0.5	−2.0	
Initial angular speed (rad/s) bat	2.0	2.0	2.0	
Final x-velocity (m/s) ball	_____	_____	_____	
Final y-position (m) ball	−3.0	−0.5	−2.0	
Final angular speed (rad/s) bat	−2.0	1.2	_____	

$$\text{SUMMARY}$$

- Angular Momentum of Point Particle: $L_z = x\, p_y - y\, p_x = x\, (m\, v_y) - y\, (m\, v_x).$
- Angular Momentum of Extended Body: $L_z = I\,\omega.$
- Conservation of Momentum Principle: ΣL_z (before collision) = ΣL_z (after collision)

For the ball-beam system:

$L_{zo}(\text{beam}) + L_{zo}(\text{ball}) = L_z(\text{beam}) + L_z(\text{ball})$

Observe the behavior of the negatively charged ball for Question 1. Then complete the following information.

Question 2: With v_{xo} = +1.50 m/s and y_o = +2.00 m, determine the initial angular momentum of the 1.0-kg ball.

Question 3: Run the simulation with v_{xo} = +1.50 m/s and y_o = +2.00 m. Stop the motion in two quadrants and use the x, y, v_x, and v_y meter readings to calculate the angular momentum of the 1.0-kg ball when it is at these positions. Compare these values to the initial angular momentum calculated in Question 2.

Questions 4-5: Why is angular momentum conserved?

Newton's Second Law can be written in the following form that applies to rotational motion:

$$\text{Net torque} = \text{Change in angular momentum}$$
$$\Sigma \tau = dL / dt$$

Note that if the net torque on an object is zero, its angular momentum is constant. Show that the net torque on the moving particle in this simulation is zero.

Show forces on moving particle: Determine torque caused by these forces:

SUMMARY

- Angular Momentum of Point Particle: $L_z = x\, p_y - y\, p_x = x\, (m\, v_y) - y\, (m\, v_x)$.

- Angular Momentum of Extended Body: $L_z = I\, \omega$.

- Conservation of Momentum Principle: If the net torque on a system is zero, the system's angular momentum is constant.

8

THERMODYNAMICS

Thermodynamics Review Notes:

- Ideal gas law: . . . . $PV = NkT = nRT$.
. . . .
- W is the work done *by* the gas on the environment: . . . . $W = \int_{vo}^{v} P\, dV.$
. . . .
- First Law of Thermodynamics: . . . . $Q = W + U_{in}.$

- Internal energy of an ideal gas: . . . . $U_{in} = (3/2)NkT = (3/2)nRT.$

Units

1.0 dm = 0.10 m

$1.0\ (dm)^3 = 1.0 \times 10^{-3}\ m^3$

$1\ kPa = 1.0 \times 10^3\ Pa = 1.0 \times 10^3\ N/m^2$

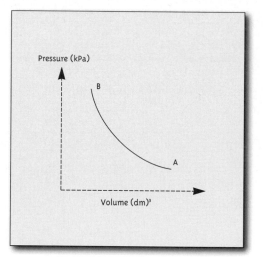

Question 1 — Atom's Speed: Does the colored atom in the simulation always move at the same speed, or does the speed change irregularly?

Question 2 — Reason: What causes the speed of an atom to change irregularly?

Question 3 — Type of Collision with Wall: In this simulation, is an atom's collision with the wall more like an elastic collision or an inelastic collision? Explain.

Question 4 — Instantaneous Kinetic Energy: At one instant of time, do all of the atoms move with the same speed and hence the same average kinetic energy?

Question 5 — Average Speed and Temperature: Set the temperature of the gas to 100 K and make a rough estimate of the average speed of the atom during a one-minute interval. Then predict its average speed at 400 K. After your prediction, set the thermometer to 400 K and estimate the average speed of the colored atom during a one-minute interval. Does the temperature seem to be proportional to the average speed or to the average kinetic energy of the atom?

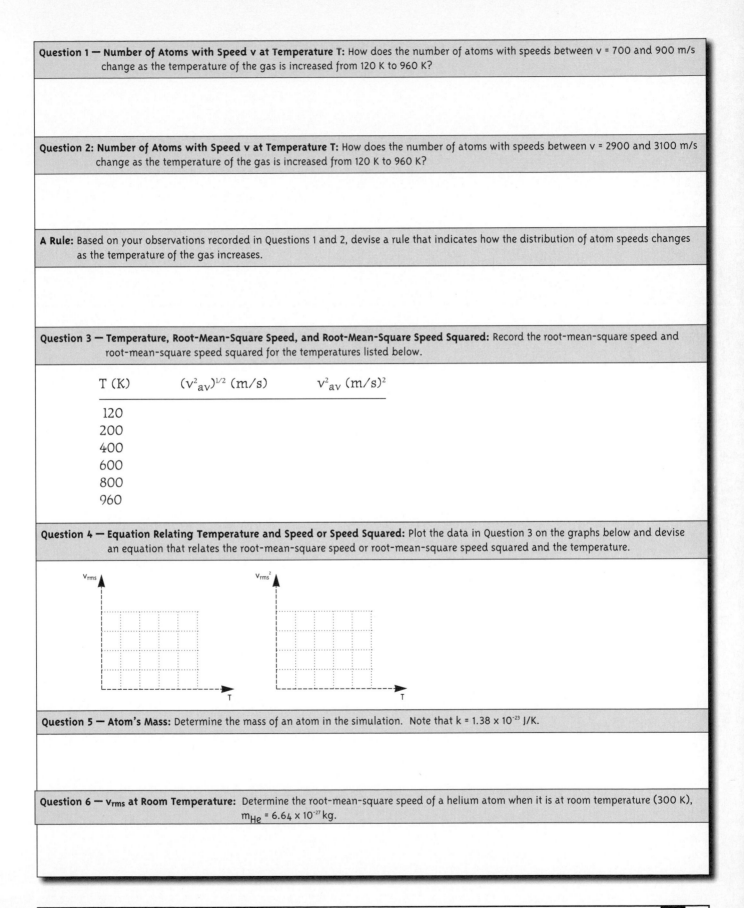

Question 1 — Number of Atoms with Speed v at Temperature T: How does the number of atoms with speeds between v = 700 and 900 m/s change as the temperature of the gas is increased from 120 K to 960 K?

Question 2: Number of Atoms with Speed v at Temperature T: How does the number of atoms with speeds between v = 2900 and 3100 m/s change as the temperature of the gas is increased from 120 K to 960 K?

A Rule: Based on your observations recorded in Questions 1 and 2, devise a rule that indicates how the distribution of atom speeds changes as the temperature of the gas increases.

Question 3 — Temperature, Root-Mean-Square Speed, and Root-Mean-Square Speed Squared: Record the root-mean-square speed and root-mean-square speed squared for the temperatures listed below.

T (K)	$(v^2_{av})^{1/2}$ (m/s)	v^2_{av} (m/s)^2
120		
200		
400		
600		
800		
960		

Question 4 — Equation Relating Temperature and Speed or Speed Squared: Plot the data in Question 3 on the graphs below and devise an equation that relates the root-mean-square speed or root-mean-square speed squared and the temperature.

Question 5 — Atom's Mass: Determine the mass of an atom in the simulation. Note that $k = 1.38 \times 10^{-23}$ J/K.

Question 6 — v_{rms} at Room Temperature: Determine the root-mean-square speed of a helium atom when it is at room temperature (300 K), $m_{He} = 6.64 \times 10^{-27}$ kg.

Question 1: — Root-Mean-Square Speed of a Helium Atom: Determine the root-mean-square speed of a helium atom in a gas at temperature 300 K and 600 K. The mass of a helium atom is 6.64×10^{-27} kg, and the Boltzmann constant k is 1.38×10^{-23} J/K.

Question 2 — Most Probable Speed of a Helium Atom: Determine the most probable speed of a helium atom in a gas at temperature 300 K and at 600 K. The mass of a helium atom is 6.64×10^{-27} kg, and the Boltzmann constant k is 1.38×10^{-23} J/K.

Question 3 — Average Random Kinetic Energy of Atom in Gas: Determine the average random kinetic energy of an atom in a gas that is at temperature 300 K (approximately room temperature). Convert the energy to electron volts (1 eV = 1.6×10^{-19} J).

Question 4 — Can an Atom Break a Bond? Compare the average random kinetic energy of an atom in a gas at room temperature to the energy needed to break a bond in a DNA molecule or knock a hydrogen atom off of a protein. Is the average kinetic energy important, or is some other property of the gas important? If so, describe that property.

Question 5 — Speed to Break a Bond in DNA: Determine the speed at which a helium atom would have to move in order to have enough energy to break a bond in a DNA molecule.

Questions 1-3: Draw a graph that has the general shape of each process below. The graph must be consistent with the ideal gas law (PV = nRT).

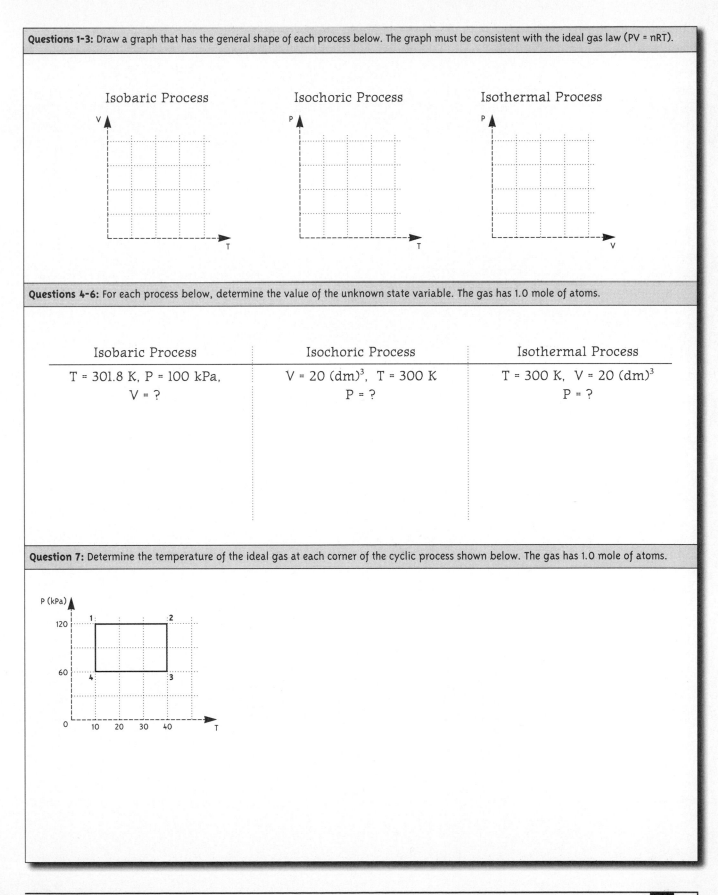

Isobaric Process

Isochoric Process

Isothermal Process

Questions 4-6: For each process below, determine the value of the unknown state variable. The gas has 1.0 mole of atoms.

Isobaric Process	Isochoric Process	Isothermal Process
T = 301.8 K, P = 100 kPa, V = ?	V = 20 $(dm)^3$, T = 300 K P = ?	T = 300 K, V = 20 $(dm)^3$ P = ?

Question 7: Determine the temperature of the ideal gas at each corner of the cyclic process shown below. The gas has 1.0 mole of atoms.

Pressure (kPa)

Volume (dm)^3

Question 1: On the simulation for Question 1, pull the top of the left vertical bar on the graph and adjust it to 20 (dm)^3 and 100 kPa. Set the top of the right bar to 40 (dm)^3 and 100 k Pa. Then calculate the work done by the gas in moving from the first set of values to the second.

Question 2: Adjust the top of the right bar to volume 40 (dm)^3 and pressure 200 kPa. (The left bar remains as in Question 1.) Predict the work that will be done by the gas as it moves in a straight line from volume 20 (dm)^3 and pressure 100 kPa to volume 40 (dm)^3 and pressure 200 kPa. Then, run the simulation to check your answer.

Question 3: The simulation for Question 3 shows an isothermal process in which the pressure is related to the volume by the equation P = (2500 N.m)/V. Determine the work done by the gas as it moves from volume 40 (dm)^3 to volume 10 (dm)^3.

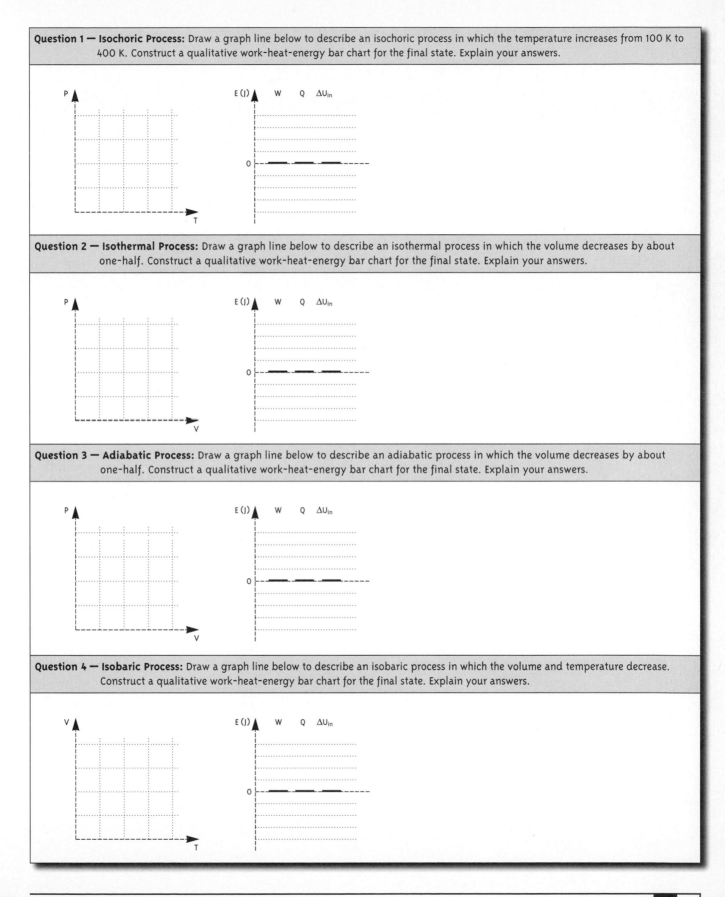

Question 1 — Isochoric Process: Draw a graph line below to describe an isochoric process in which the temperature increases from 100 K to 400 K. Construct a qualitative work-heat-energy bar chart for the final state. Explain your answers.

Question 2 — Isothermal Process: Draw a graph line below to describe an isothermal process in which the volume decreases by about one-half. Construct a qualitative work-heat-energy bar chart for the final state. Explain your answers.

Question 3 — Adiabatic Process: Draw a graph line below to describe an adiabatic process in which the volume decreases by about one-half. Construct a qualitative work-heat-energy bar chart for the final state. Explain your answers.

Question 4 — Isobaric Process: Draw a graph line below to describe an isobaric process in which the volume and temperature decrease. Construct a qualitative work-heat-energy bar chart for the final state. Explain your answers.

Question 1 — First Law Bar Chart: The bar chart below represents the First Law of Thermodynamics description of heat transfer to a system that remains at constant volume. Why does the bar chart have this shape?

Question 2 — Constant-Volume Heat Capacity ($\approx$): By taking incremental increases in temperature on the simulation, estimate the constant-volume heat capacity. The gas has 1.0 mole of atoms.

$$C_V = \frac{1}{n} \frac{dQ}{dT}$$

at constant volume

Question 3 — Constant-Volume Heat Capacity: Use the simulation numbers as the gas moves at constant volume from temperature 200 K to 800 K to calculate its constant-volume heat capacity. The gas has 1.0 mole of atoms.

$$C_V = \frac{1}{n} \frac{Q}{T-T_0}$$

at constant volume

Question 4 — Expression for Cv: Use the following principles to show that the constant-volume heat capacity of an ideal gas is given by the expression below. Start with the equation in Question 3.

• Use the First Law of Thermodynamics.
• Consider internal energy of a gas in terms of temperature.
• Rearrange to get the expression at the right.

$$C_V = \frac{3}{2} R$$

at constant volume

Question 5 — First Law Bar Chart: The bar chart below represents the First Law of Thermodynamics description of heat transfer to a system that remains at constant pressure. Why does the bar chart have this shape?

Question 6 — Constant-Pressure Heat Capacity ($\approx$): By taking incremental increases in temperature on the simulation, estimate the constant-pressure heat capacity. The gas has 1.0 mole of atoms.

$$C_p = \frac{1}{n}\frac{dQ}{dT}$$

at constant pressure

Question 7 — Constant-Pressure Heat Capacity: Use the simulation numbers as the gas moves at constant pressure from temperature 200 K to 800 K to calculate its constant-pressure heat capacity. The gas has 1.0 mole of atoms.

$$C_p = \frac{1}{n}\frac{Q}{T-T_0}$$

at constant pressure

Question 8 — Derive Expression for Cp: Use the following principles to show that the constant-pressure heat capacity of an ideal gas is given by the expression below. Start with the equation in Question 7.

• Use the First Law of Thermodynamics.
• Consider internal energy of a gas in terms of temperature.
• Consider work done by the gas at constant pressure.
• Use the ideal gas law.
• Rearrange to get the expression at the right.

$$C_p = \frac{5}{2}R$$

at constant pressure

The piston encloses the gas so that the cylinder volume is 13.40 (dm)³. Run the simulation and watch the process. The gas starts at 21 kPa and 100 K, moves to 100 kPa and 495 K, and then returns to where it started. Answer the questions below and on the next page.

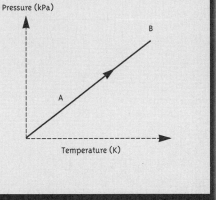

Question 1: What happens to the gas during the first part of the process as the system moves from A to B?

(a) Does the volume increase, remain the same, or decrease?

(b) Does the pressure increase, remain the same, or decrease?

(c) Does the temperature increase, remain the same, or decrease?

Predict the direction of change of bars in the bar chart as the gas moves from A to B:

Q (heat transferred to [+] or from [-] the system), W (work done by the system [+] or on the system [-]), and U (internal energy change of the system).

(d) Q: negative   no change   positive

(e) W: negative   no change   positive

(f) U: negative   no change   positive

After you make your predictions, run the simulation to check them.

Question 2: Use the ideal gas law to determine the number of moles of gas in the system and the number of atoms in the system.

Question 3 — Pressure at B: Determine the pressure in the gas when it is at B. The volume of the gas is 40 (dm)³, and the temperature is 495 K.

Question 4 — Internal Energy Change: Determine the internal energy change of the gas as it moves from A to B. Use the values of the state variables determined in Questions 2 and 3.

Question 5 — Heat Transfer: Use the First Law of Thermodynamics to determine the heat transfer to the gas as it moves from A to B. Use the results from previous questions.

Question 6: Use a similar procedure to determine the heat transfer to the gas for the following isochoric process.

- The volume remains constant at 20.9 (dm)³.
- The pressure changes from 40 kPa to 197 kPa.
- The temperature starts at 100 K and ends at some unknown temperature.

Watch the piston and graph in the isobaric process simulation, The pressure is set at 120 kPa. The initial volume and temperature at A are 41.6 (dm)^3 and 600 K, respectively. Answer the questions below and on the next page.

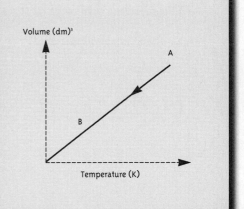

Question 1: What happens to the gas during the first part of the process as the system moves from A to B?

(a) Does the volume increase, remain the same, or decrease?

(b) Does the pressure increase, remain the same, or decrease?

(c) Does the temperature increase, remain the same, or decrease?

Predict the direction of change of bars in the bar chart as the gas moves from A to B:

Q (heat transferred to [+] or from [−] the system), W (work done by the system [+] or on the system [−]), and U (internal energy change of the system).

(d) Q: negative   no change   positive

(e) W: negative   no change   positive

(f) U: negative   no change   positive

After you make your predictions, run the simulation to check them.

Question 2: Use the ideal gas law to determine the volume of the gas when it is at B. The temperature at B is 150 K, and the pressure remains constant at 120 kPa.

Question 3 — Work Done By Gas: Determine the work done by the gas as the gas moves from A to B. The pressure remains constant at 120 kPa. The volume changes from 41.6 (dm)³ to 10.4 (dm)³, The temperature changes from 600 K to 150 K. (Note that the graph at the right is *not* a pressure-versus-volume graph.)

Question 4 — Internal Energy Change: Determine the internal energy change of the gas as it moves from A to B. Use the values of the state variables determined in Question 2.

Question 5 — Heat Transfer: Use the First Law of Thermodynamics to determine the heat transfer to the gas as it moves from A to B. Use the results from previous questions.

Question 6 — Use a similar procedure to determine the heat transfer for the following isobaric process.

- The pressure remains constant at 100 kPa.
- The volume changes from 41.6 (dm)³ to 10.4 (dm)³.
- The temperature starts at 500 K and ends at some unknown temperature.

Watch the piston and graph in the isothermal process simulation. Set the temperature to 300 K. The initial volume and pressure at A are 40 (dm)³ and 62 kPa, respectively. Answer the questions on this page and the next.

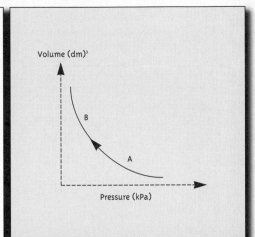

Question 1: What happens to the gas during the first part of the process, as the system moves from A to B?

(a) Does the volume increase, remain the same, or decrease?

(b) Does the pressure increase, remain the same, or decrease?

(c) Does the temperature increase, remain the same, or decrease?

Predict the direction of change of bars in the bar chart as the gas moves from A to B:

Q (heat transferred to [+] or from [-] the system), W (work done by the system [+] or on the system [-]), and U (internal energy change of the system).

(d) Q: negative   no change   positive

(e) W: negative   no change   positive

(f) U: negative   no change   positive

After you make your predictions, run the simulation to check them.

Question 2: Use the ideal gas law to determine the pressure in the gas when it is at B. The volume at B is 10 (dm)³, and the temperature remains constant at 300 K.

Question 3 — Work Done By Gas: Determine the work done by the gas as the gas moves from A to B. The temperature remains constant at 300 K. The volume changes from 40 (dm)^3 to 10.0 (dm)^3, The pressure changes from 62 kPa to 246 kPa. Note: You will have to use the ideal gas law to get the work integral in an integrable form.

Volume (dm)^3

B

A

Pressure (kPa)

Question 4 — Heat Transfer: Use the First Law of Thermodynamics to determine the heat transfer to the gas as it moves from A to B. Use the results from previous questions.

Question 5: What happens to the graph line if the temperature is decreased from 350 K to 250 K?

(a) rises but retains shape

(b) stays the same

(c) drops but retains shape

Explain your answer.

Question 6: Use a similar procedure to determine the heat transfer for the following isothermal process.

- The temperature remains constant at 250 K.
- The volume changes from 40 (dm)^3 to 10 (dm)^3.
- The pressure starts at 52 kPa and ends at some unknown pressure.

Observe the motion of the atoms in the Adiabatic process simulation. Adjust the initial temperature to 200 K. The initial volume and pressure are 40 (dm)³ and 42 kPa, respectively. Answer the questions below and on the next page.

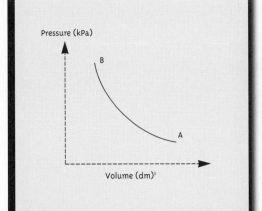

Question 1: What happens to the gas during the first part of the process, as the system moves from A to B?

(a) Does the volume increase, remain the same, or decrease?

(b) Does the pressure increase, remain the same, or decrease?

(c) Does the temperature increase, remain the same, or decrease?

Predict the direction of change of bars in the bar chart as the gas moves from A to B:
Q (heat transferred to [+] or from [−] the system), W (work done by the system [+] or on the system [−]), and U (internal energy change of the system).

(d) Q: negative   no change   positive

(e) W: negative   no change   positive

(f) U: negative   no change   positive

Justify each choice in the space below.

Question 2: Determine the temperature of the gas when it is at B. The volume of the 1.0-mole of gas is 10 (dm)³, and the pressure is 419 kPa.

Question 3 — Internal Energy Change: Determine the change in the internal energy of the gas as it moves from A to B.

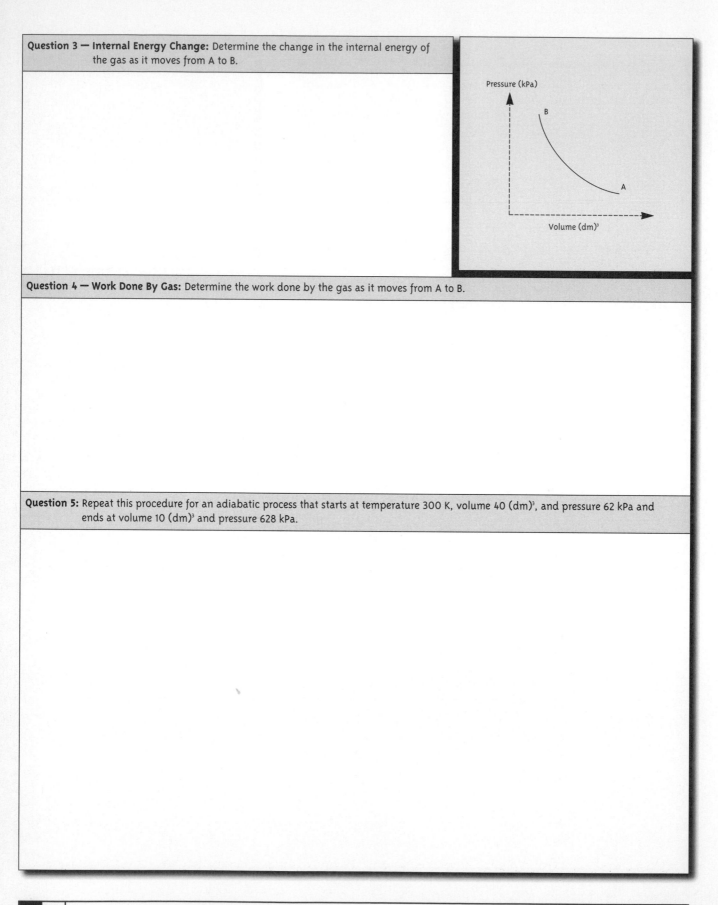

Pressure (kPa)

B

A

Volume (dm)^3

Question 4 — Work Done By Gas: Determine the work done by the gas as it moves from A to B.

Question 5: Repeat this procedure for an adiabatic process that starts at temperature 300 K, volume 40 (dm)^3, and pressure 62 kPa and ends at volume 10 (dm)^3 and pressure 628 kPa.

Question 1 — Temperature: Use the ideal gas law (PV = nRT) to determine the temperature of the gas at each corner in the process represented by the pressure-versus-volume (PV) graph below.

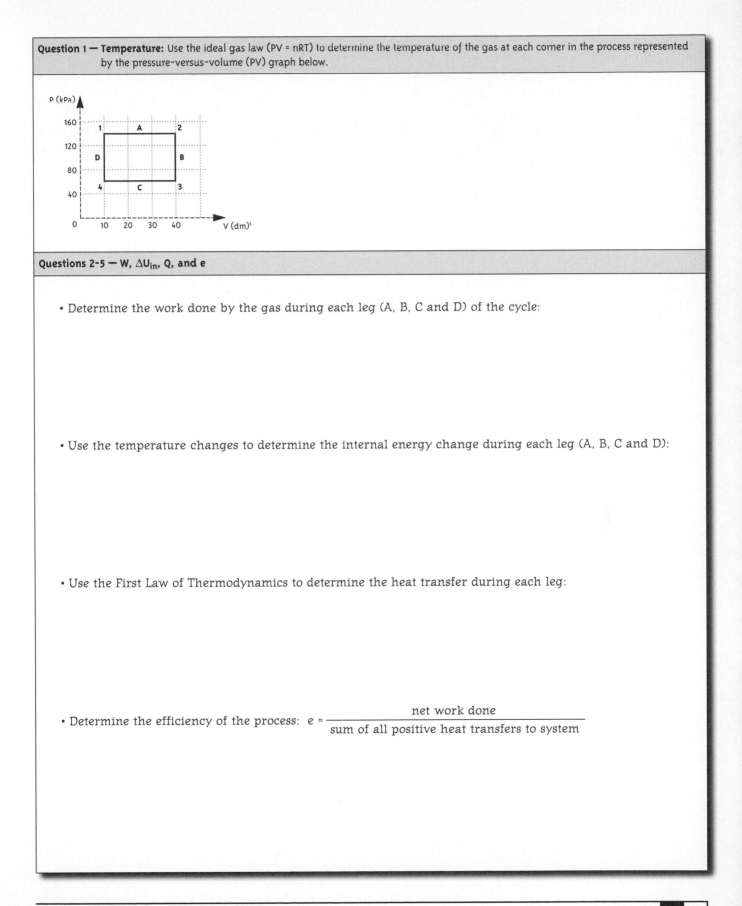

Questions 2-5 — W, ΔU_{in}, Q, and e

- Determine the work done by the gas during each leg (A, B, C and D) of the cycle:

- Use the temperature changes to determine the internal energy change during each leg (A, B, C and D):

- Use the First Law of Thermodynamics to determine the heat transfer during each leg:

- Determine the efficiency of the process: $e = \dfrac{\text{net work done}}{\text{sum of all positive heat transfers to system}}$

Question 1 — Temperature: Use the ideal gas law (PV = nRT) to determine the temperature of the gas at each corner in the process represented by the pressure-versus-volume (PV) graph below.

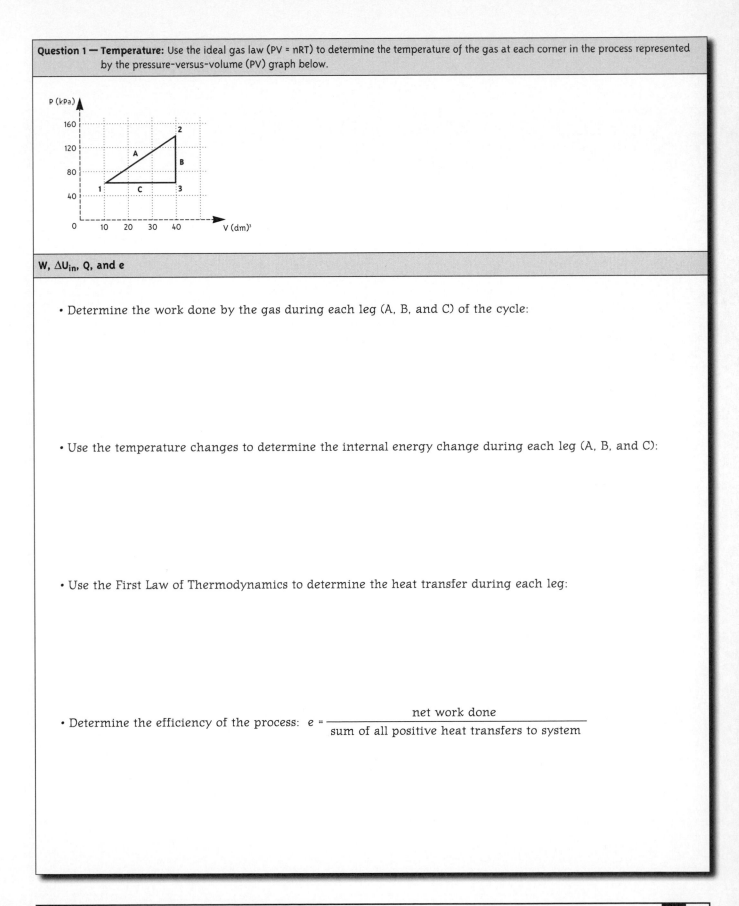

W, ΔU_{in}, Q, and e

• Determine the work done by the gas during each leg (A, B, and C) of the cycle:

• Use the temperature changes to determine the internal energy change during each leg (A, B, and C):

• Use the First Law of Thermodynamics to determine the heat transfer during each leg:

• Determine the efficiency of the process: $e = \dfrac{\text{net work done}}{\text{sum of all positive heat transfers to system}}$

Question 2 — Temperature: Use the ideal gas law (PV = nRT) to determine the temperature of the gas at each corner in the process represented by the pressure-versus-volume (PV) graph below.

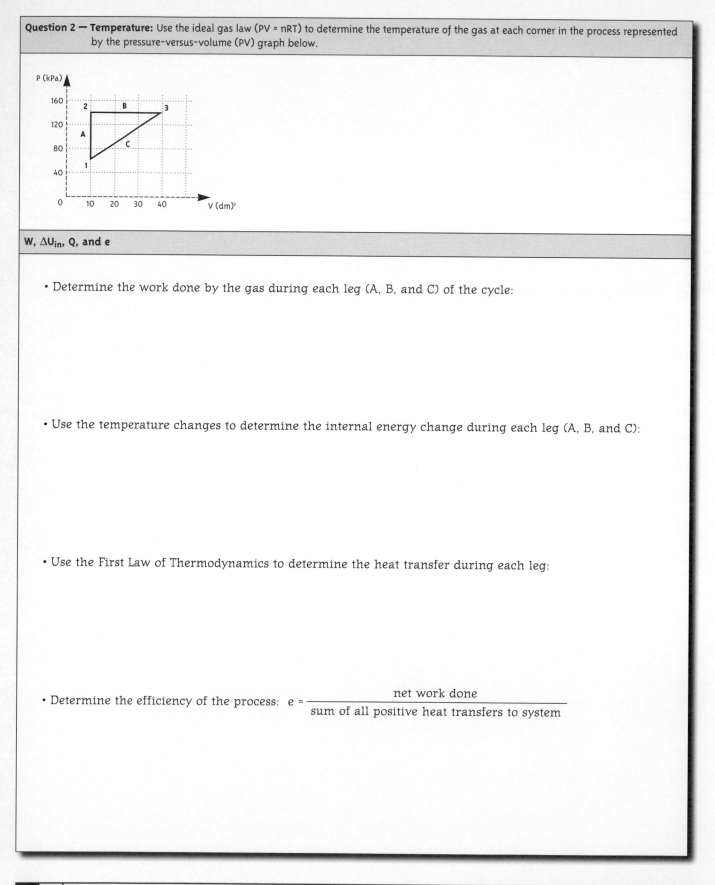

W, ΔU_{in}, Q, and e

- Determine the work done by the gas during each leg (A, B, and C) of the cycle:

- Use the temperature changes to determine the internal energy change during each leg (A, B, and C):

- Use the First Law of Thermodynamics to determine the heat transfer during each leg:

- Determine the efficiency of the process: $e = \dfrac{\text{net work done}}{\text{sum of all positive heat transfers to system}}$

Question 3 — Temperature: Use the ideal gas law (PV = nRT) to determine the temperature of the gas at each corner in the process represented by the pressure-versus-volume (PV) graph below.

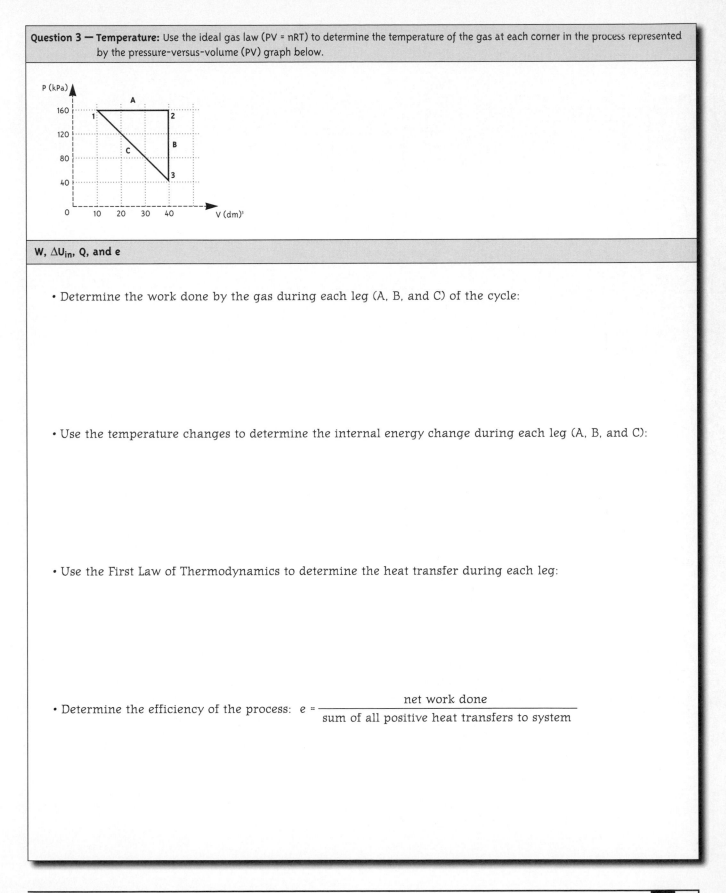

W, ΔU_{in}, Q, and e

- Determine the work done by the gas during each leg (A, B, and C) of the cycle:

- Use the temperature changes to determine the internal energy change during each leg (A, B, and C):

- Use the First Law of Thermodynamics to determine the heat transfer during each leg:

- Determine the efficiency of the process: $e = \dfrac{\text{net work done}}{\text{sum of all positive heat transfers to system}}$

Question 4: Apply the strategies used in the previous cyclic-process problems to determine the efficiencies of the two processes represented on the pressure-versus-volume (PV) graphs below.

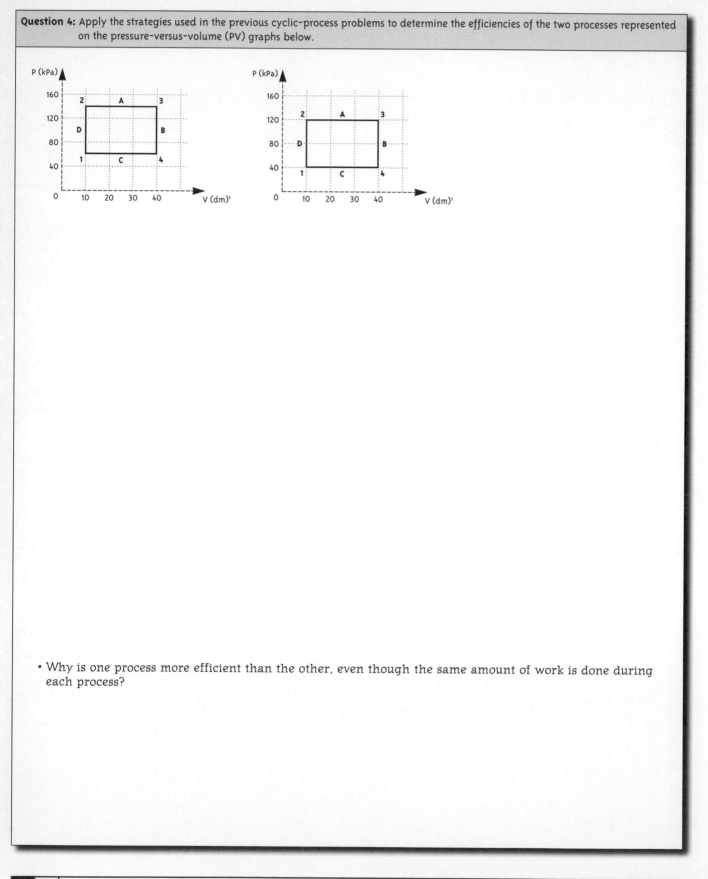

• Why is one process more efficient than the other, even though the same amount of work is done during each process?

Question 1 — U_{in}, W, and Q Changes: Predict the signs of the changes in the internal energy, work done by the system, and heat transfer to the system for each stage in the Carnot cycle. A and C are isothermal processes, and B and D are adiabatic processes.

	ΔU_{in}	W	Q
A	+ 0 −	+ 0 −	+ 0 −
B	+ 0 −	+ 0 −	+ 0 −
C	+ 0 −	+ 0 −	+ 0 −
D	+ 0 −	+ 0 −	+ 0 −

Question 2 — Pressure at Corners: Determine the pressure at each corner. The temperature and volume at the corners are given.

	T (K)	V (dm)^3	P (kPa)
1	500	10	
2	500	19	
3	300	40	
4	300	21	

Question 3 — U_{in}, W, and Q: Determine the change in the system's internal energy, the work done by the gas, and the heat transfer to the gas for each leg of the process.

	ΔU_{in}	W	Q
A			
B			
C			
D			

Question 4 — Efficiency: Use the numbers above to determine the efficiency e of the process (the ratio of the net work done and the heat input to the system).

Question 5 — Efficiency: Compare the above calculation of efficiency to that obtained using a special equation for efficiency derived from the first and second laws of thermodynamics: $e = 1 - T_{hot}/T_{cool}$, where T_{hot} is the temperature of the hot reservoir (500 K for this example), and T_{cool} is the temperature of the cool reservoir (300 K for this example).

9

VIBRATIONS

Question 1: Write the amplitude of each block.

Top Block:

Bottom Block:

Question 2: Write the frequency of each block.

Top Block:

Bottom Block:

Question 3: Write the phase angle of each block.

Top Block:

Middle Block

Bottom Block:

Question 4: Write an equation that describes the position of the top block as a function of time, and another equation that describes the position of the bottom block.

Top Block:

Bottom Block:

Question 5: Noting that v = dx/dt, write an equation that describes the velocity of the top block as a function of time, and another equation that describes the velocity of the bottom block.

Top Block:

Bottom Block:

Question 6: Noting that a = dv/dt, write an equation that describes the acceleration of the top block as a function of time, and another equation that describes the acceleration of the bottom block.

Top Block:

Bottom Block:

Question 1: Observe the motion of the block and construct a position-versus-time graph (x-vs-t) for the motion. Be sure to include a scale with the correct units.

Question 2: Suppose that the equation $x = x_m \cos (2\pi f t)$ describes this type of motion and the above graph. Determine the amplitude x_m, the period T, and the frequency f of this particular motion. Insert any of these values into this equation to get a specific equation that describes this motion. Check to see if the equation gives the correct values of x for different specific times.

Question 3: Use the equation for x in Question 2 and the fact that velocity is the rate of change of position ($v = dx/dt$) to determine an equation for the block's velocity as a function of time. Be sure to include an estimate of the block's maximum speed. Compare the value of v given by your equation to the value of v given in the simulation for the five times listed.

Question 4: Run the simulation that shows both x-vs-t and v-vs-t graphs at the same time. Does the value of v at every time seem to equal qualitatively the slope of the x-vs-t graph at that time?

Question 5: Use the equation developed in Question 3 and the fact that the acceleration is the rate of change of the velocity ($a = dv/dt$) to determine an equation for this block's acceleration as a function of time. Be sure to include an estimate of the block's maximum acceleration a_m. Compare the value of the acceleration given by your equation and from the simulation graph for several specific times.

Question 6: Acceleration is the time derivative of the velocity ($a = dv/dt$). This derivative is also the slope of the velocity-versus-time graph. Move the slider back and forth to compare slopes and accelerations at specific times.

• The slope of the v-vs-t graph and the acceleration at times 1.0 s, 2.0 s, and 3.0 s

• The slope of the v-vs-t graph and the acceleration at times 0.5 s and 2.5 s

• The slope of the v-vs-t graph and the acceleration at times 1.5 s and 3.5 s

Summary: Draw below the x-vs-t, v-vs-t, and a-vs-t graphs described in this activity. Indicate roughly the vertical scale for each graph. Note that the slopes of the graphs are consistent with the definitions of velocity and acceleration.

Question 1: Determine two times when the block's kinetic energy and velocity are greatest.

Question 2: Determine two times when the block's elastic potential energy is greatest.

Question 3: The block's mass in this simulation is 2.0 kg, and the force constant of the spring is 40 N/m. Note the block's speed at different times and determine the amplitude of the vibration.

Question 4: The block's mass in this simulation is 4.0 kg. Note the meter readings and determine the force constant of the spring.

Question 5: The block's mass in this simulation is 2.0 kg, the spring's force constant is 40 N/m, and the vibrational amplitude is 1.6 m. Determine the speed of the block when the spring is displaced 0.90 m from equilibrium.

Tarzan Jr. is on a spring, ready to be weighed. Based on your careful observations of the simulation, use a frequency approach and then an energy approach to independently measure Tarzan Jr.'s mass. The spring force constant is 200 N/m, and the gravitational constant is 10 N/kg.

Question 1 — Frequency Approach:

Question 2 — Energy Approach:

A 20-kg ape holds the 40-kg aging Tarzan as they vibrate up and down at the end of a springlike vine. At the lowest position in the vibration, after vibrating for 1.5 s, the ape accidentally drops Tarzan.

Question 1: How does the ape's maximum acceleration change?

   (a) increases   (b) remains the same   (c) decreases

Explain the reason for your choice:

Question 2: How does the ape's maximum velocity change?

   (a) increases   (b) remains the same   (c) decreases

Explain the reason for your choice:

A 90-kg skier holds a 30-kg cart that vibrates at the end of a spring. After one vibration, the skier lets go of the cart when it is at its maximum displacement from equilibrium.

Question 1: After the skier's release, how does the cart's new maximum displacement from equilibrium compare to before?

___ 1/4   ___ 1/2   ___ same   ___ 2 times   ___ 4 times

Explain.

Question 2: After the skier's release, how does the cart's new frequency compare to before?

___ 1/4   ___ 1/2   ___ same   ___ 2 times   ___ 4 times

Explain.

Question 3: After the skier's release, how does the cart's new maximum speed compare to before?

___ 1/4   ___ 1/2   ___ same   ___ 2 times   ___ 4 times

Explain.

A 60-kg skier holds on to a 20-kg cart that vibrates at the end of a spring. After 1.5 vibrations, the skier lets go of the cart when the cart is passing through its equilibrium position at maximum speed.

Question 1: After the skier's release, how does the cart's new maximum displacement from equilibrium compare to before?

___ 1/4   ___ 1/2   ___ same   ___ 2 times   ___ 4 times

Explain.

Question 2: After the skier's release, how does the cart's new frequency compare to before?

___ 1/4   ___ 1/2   ___ same   ___ 2 times   ___ 4 times

Explain.

Question 3: After the skier's release, how does the cart's new maximum speed compare to before?

___ 1/4   ___ 1/2   ___ same   ___ 2 times   ___ 4 times

Explain.

Question 1 — Force Constant: A 1.0-kg block vibrates at the end of a spring. Based on your observations, determine the force constant of the spring.

Question 1 — Speed: Now, based on your observations and previous calculations, determine the maximum speed of the block.

Question 2: A 1.0-kg block is attached to 3.0-N/m springs on each side. The springs are initially relaxed, and the block is moving right at speed 8.0 m/s. Determine the time interval needed for one vibration and the frequency of vibration. Note that you will have to determine the effective force constant of the two-spring system. Construct a force diagram for the block and apply Newton's Second Law when it is displaced a distance x from equilibrium. Compare this equation to the equation for a block at the end of a single spring.

(a) Calculate the force constant of the spring attached to the 510-g cart. The incline is 168 cm long and the height of the end of the incline is 65 cm. You may need to use other quantities shown in the video. (b) Then, estimate the coefficient of friction between the cart and incline. (c) Finally, predict the frequency of a similar system with two springs attached in parallel to the cart.

Concept(s) to be used:

Known or estimated quantities:

Unknown to be determined:

Calculations:

A 50-kg skier, after coming down the slope, travels at 12 m/s on a level surface, raises a ski pole, and runs into a 25-kg cart. The ski pole attaches to the vibrating cart. A spring on the other side of the cart compresses and expands, giving the skier a Vibro-Ride at the end of the ski run—what fun! Determine the amplitude of the vibration and the period for one vibration.

Questions 1-3 — Amplitude of Vibration:

• Plan a strategy for determining the amplitude of the vibration.

• Solve the problem parts and determine the vibration amplitude.

Question 4 — Period for One Vibration:

Question 1: How does the frequency of a pendulum depend on the mass of the pendulum bob?

___ Frequency is lower for a larger mass bob.

___ Frequency is higher for a larger mass bob.

___ Frequency is independent of the mass of the bob.

Explain your answer.

Question 2: How does the frequency of a pendulum depend on the amplitude of the vibration?

___ Frequency is lower for a larger amplitude vibration.

___ Frequency is higher for a larger amplitude vibration.

___ Frequency is independent of the amplitude of the vibration.

Explain your answer.

Question 3: How does the frequency of a pendulum depend on the length of the string?

___ Frequency is lower for a longer length pendulum string.

___ Frequency is higher for a longer length pendulum string.

___ Frequency is independent of the length of the pendulum string.

Explain your answer.

Three different length pendula hang from a metal rod. Harold can move the rod back and forth and cause only one pendulum to have large amplitude vibrations. How can he do this?

Explanation:

Question 1: As a visiting scientist on planet Zeus, you are asked to assist in the design of a pendulum timer for use on the planet. The gravitational constant on Zeus is 6.0 N/kg = 6.0 m/s^2. The pendulum bob you will be using has a mass of 0.10 kg. You are to choose the length of the pendulum string so that the bob completes one vibration in 3.0 s. Complete your calculations. Then adjust the length of the string and run the simulation to check your work.

Question 1: You must decide the constant speed to walk in order to make it under the 4.0-m-long pendulum with its 40-N bob. If you choose the wrong speed, you get a good bump on the head — or somewhere else. The pendulum starts at rest in the position shown. The grid lines in the simulation are separated by 1.0 m. The gravitational constant is 10 N/kg.

Solution:

First construct a free-body diagram for the physical pendulum and then apply the rotational form of Newton's Second Law to its motion. This leads to a differential equation with a standard solution. Finally, examine the solution to determine the frequency of the motion. Then solve for the length of the string of the simple pendulum that has the same frequency.

Free-Body Diagram: Construct a free-body diagram for the hanging uniform rod when it is displaced from its equilibrium position.

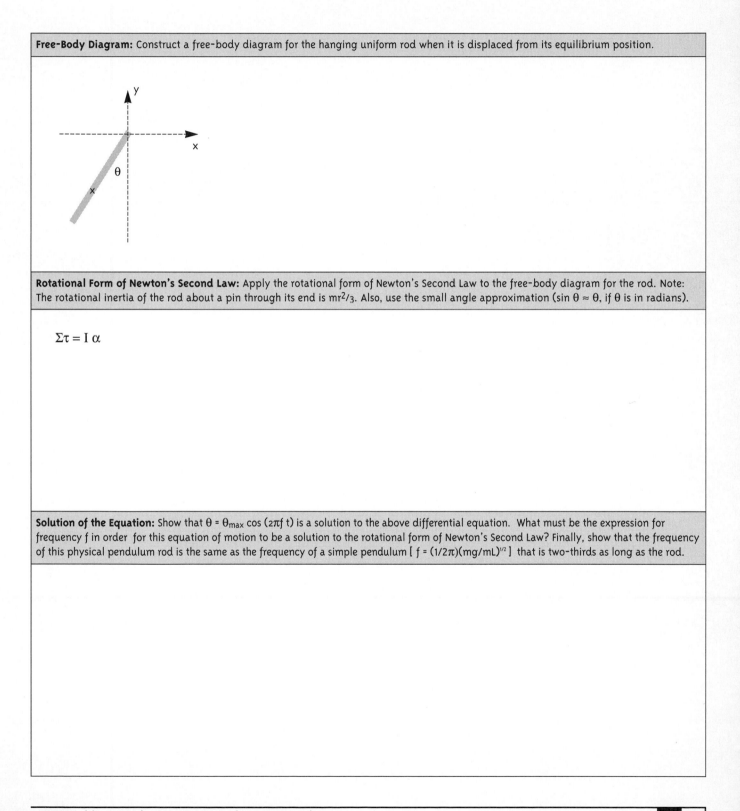

Rotational Form of Newton's Second Law: Apply the rotational form of Newton's Second Law to the free-body diagram for the rod. Note: The rotational inertia of the rod about a pin through its end is $mr^2/3$. Also, use the small angle approximation ($\sin\theta \approx \theta$, if θ is in radians).

$$\Sigma\tau = I\,\alpha$$

Solution of the Equation: Show that $\theta = \theta_{max}\cos(2\pi f\,t)$ is a solution to the above differential equation. What must be the expression for frequency f in order for this equation of motion to be a solution to the rotational form of Newton's Second Law? Finally, show that the frequency of this physical pendulum rod is the same as the frequency of a simple pendulum [$f = (1/2\pi)(mg/mL)^{1/2}$] that is two-thirds as long as the rod.

A ball hits a uniform beam, or bat, at different distances from the handle. Suppose you lightly hold the handle when the ball hits the bat. Your goal is to decide where along the bat to make contact with the ball so that the bat handle makes the smallest impulse with your hands. The best spot for the ball to hit the bat is called the sweet spot.

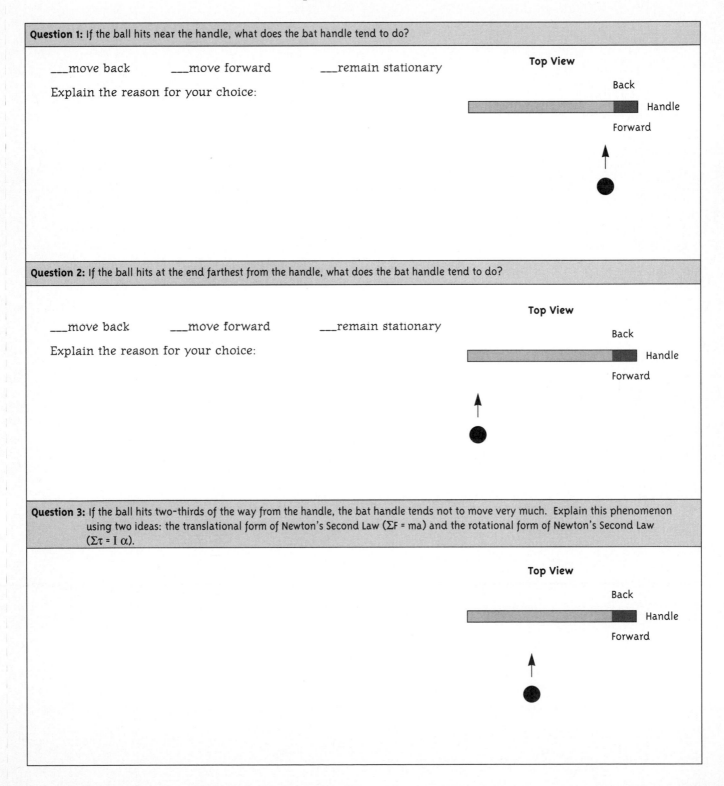

Question 1: If the ball hits near the handle, what does the bat handle tend to do?

___move back   ___move forward   ___remain stationary

Explain the reason for your choice:

Top View

Back

Handle

Forward

Question 2: If the ball hits at the end farthest from the handle, what does the bat handle tend to do?

___move back   ___move forward   ___remain stationary

Explain the reason for your choice:

Top View

Back

Handle

Forward

Question 3: If the ball hits two-thirds of the way from the handle, the bat handle tends not to move very much. Explain this phenomenon using two ideas: the translational form of Newton's Second Law ($\Sigma F = ma$) and the rotational form of Newton's Second Law ($\Sigma \tau = I \alpha$).

Top View

Back

Handle

Forward

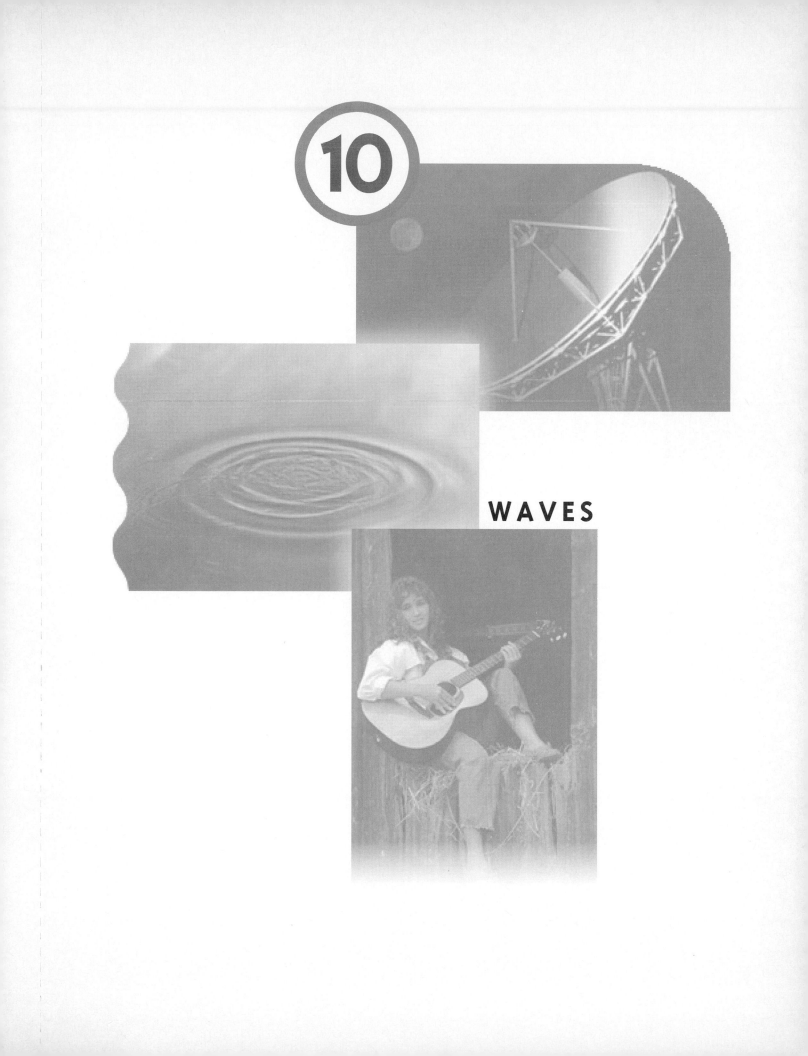

10

WAVES

Question 1 — Transverse or Longitudinal: Describe in the space below the difference between transverse waves and longitudinal waves.

Question 2 — Amplitude and Wave Speed: Does the wave amplitude affect the wave speed? Support your answer with measurements.

Question 3 — Frequency and Wave Speed: Does the wave frequency affect the wave speed? Support your answer with measurements.

Question 4 — $v = f\lambda$: Measure on the screen the wave speed and wavelength and show that $v = f\lambda$.

Question 1 — Units of $[T/\mu]^{1/2}$: Show that the units of $[T/\mu]^{1/2}$ are m/s, the same as the units of speed.

Question 2: Write an equation that describes the graph line:

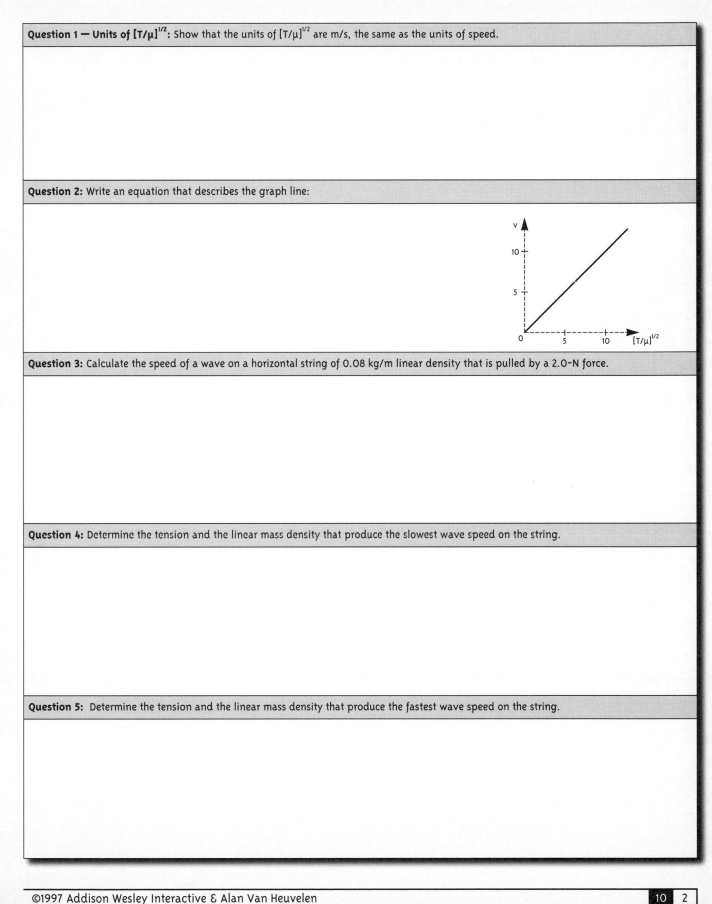

Question 3: Calculate the speed of a wave on a horizontal string of 0.08 kg/m linear density that is pulled by a 2.0-N force.

Question 4: Determine the tension and the linear mass density that produce the slowest wave speed on the string.

Question 5: Determine the tension and the linear mass density that produce the fastest wave speed on the string.

Question 1 — Units of $[B/\rho]^{1/2}$: Show that the units of $[B/\rho]^{1/2}$ are m/s, the same as the units of speed.

Question 2: Write an equation that describes the graph line.

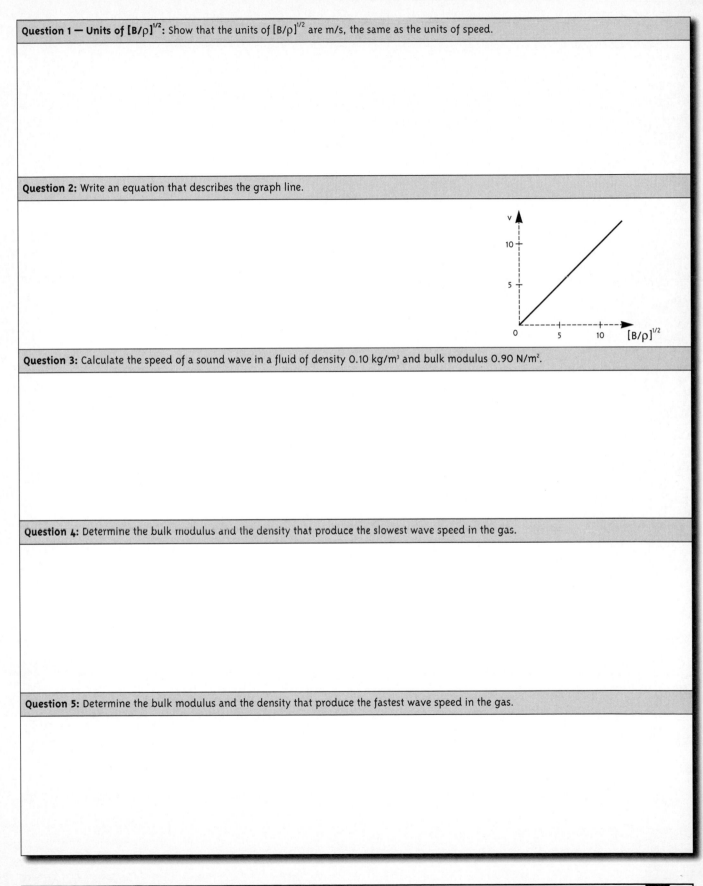

Question 3: Calculate the speed of a sound wave in a fluid of density 0.10 kg/m³ and bulk modulus 0.90 N/m².

Question 4: Determine the bulk modulus and the density that produce the slowest wave speed in the gas.

Question 5: Determine the bulk modulus and the density that produce the fastest wave speed in the gas.

Questions 5 - 7: Set the sliders on the simulation to the following values: mass/length = 30 g/m, tension = 3.0 N. Calculate the wave speed on the simulation and predict the frequencies of the first harmonic (fundamental) vibration and the second harmonic. After completing your work, adjust the frequency slider to evaluate your prediction.

First harmonic (fundamental)

Second harmonic

Follow-Up Question: Set the mass/length slider to 100 g/m and the tension to 3.6 N. Predict the frequencies of the lowest three harmonic vibrations. After completing your predictions, adjust the frequency slider to check your work.

To answer the questions, you will use the following standing-wave-on-a-string concepts:

- The speed v of a wave on a string is $v = (T / \mu)^{1/2}$.
- The allowed standing-wave wavelengths are $\lambda_n = 2L / n$ for n = 1, 2, 3, ...
- Since frequency $f = v / \lambda$, the allowed standing-wave frequencies are $f_n = (v / 2L) n$, for n = 1, 2, 3, ...

Question 1: Set the mass per unit length on the simulation to 100 g/m and the frequency to 4.0 Hz. Determine the string tension that will cause the 1.0-m-long string to vibrate at its second harmonic frequency. Check your answer by running the simulation with the predicted tension.

Question 2: After you have selected the correct tension (your answer to Question 1), the string will be vibrating in the second harmonic frequency. List below three other harmonic frequencies at which the string will vibrate. Check your predictions by adjusting the frequency slider to these frequencies.

Question 3: With the tension at 1.6 N, the mass per unit length at 100 g/m, and the frequency at 2.0 Hz, the string will vibrate in its fundamental mode (n =1). Now, double the tension to 3.2 N. How would you have to change the frequency to see a fundamental-frequency vibration again? After your prediction, adjust the simulation to your predicted frequency to check your result.

To answer the questions, you will use the following standing-wave-on-a-string concepts:

- The speed v of a wave on a string is $v = (T / \mu)^{1/2}$.
- The allowed standing-wave wavelengths are $\lambda_n = 2L / n$ for $n = 1, 2, 3, ...$
- Since frequency $f = v / \lambda$, the allowed standing-wave frequencies are $f_n = (v / 2L) n$, for $n = 1, 2, 3, ...$

Question 1: Set the mass per unit length on the simulation to 98 g/m, the tension to 1.0 N, and the frequency to 11.2 Hz. Predict the frequencies of the other six standing-wave vibrations. Check each answer by adjusting the frequency on the running simulation.

Question 2: Set the tension to 2.5 N, the mass per unit length to 40 g/m, and the frequency to the 4.0-Hz fundamental frequency. Suppose that you want another string of the same length and pulled by the same tension to vibrate at 8.0 Hz. Determine the mass per unit length needed for that string. Check your prediction by adjusting the slider in the simulation.

Question 1 — Beat Patterns: Study the graph below, which shows how pressure of sound produced at one point in space varies with time. This disturbance is caused by the combination of two sound waves of different frequencies. The resultant wave fluctuates in amplitude. Each fluctuation is called a beat, and two beats are shown. You will shortly learn how this beat pattern is formed.

Question 2: Try several frequency combinations and record the frequencies and the beat frequency.

f_1 (kHz)	f_2 (kHz)	f_{beat} (kHz)	f_1 (kHz)	f_2 (kHz)	f_{beat} (kHz)

Question 3: Compare the beat frequencies for the following two combinations:

(a) 3.4 kHz and 4.0 kHz

(b) 4.6 kHz and 4.0 kHz

Question 4: Is the beat frequency higher for frequencies of 2.0 kHz and 2.4 kHz or for frequencies of 4.0 kHz and 4.4 kHz? In short, is the beat frequency higher for higher frequency waves?

Question 5 — Beat Frequency: Based on your observations, invent a rule in the form of an equation for how the beat frequency f_{beat} is related to the frequencies f_1 and f_2 of the two waves that are producing the beats. Be sure that your rule applies to all possible combinations of the frequencies: $f_2 > f_1$, $f_2 = f_1$, and $f_2 < f_1$.

Question 8 — Does wave interference produce the beat pattern? Set both amplitudes to +30 and one frequency to 3.4 kHz and the other to 4.0 kHz. Observe that the red and green waves add together to produce large amplitude waves in the beat pattern when the red and green waves are in phase and small amplitude waves when they are not in phase. It may be easier to do this addition with the waves shown below. Add the disturbances caused by each wave at the times represented by the vertical lines. See if the resultant disturbance in the beat pattern is just the sum of the individual disturbances. When you are finished, you should be convinced that the formation of beats is just another example of wave interference.

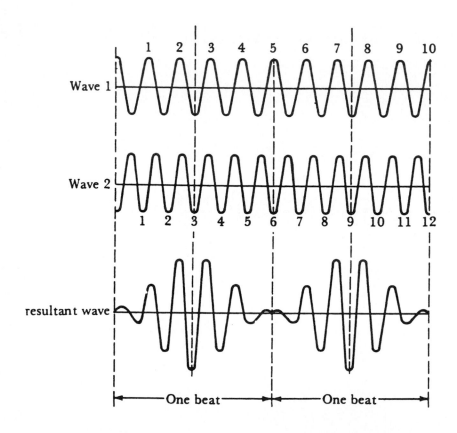

Question 9 — Complex Waves, Another Example of Wave Interference: Shown below are three wave patterns. Each pattern is the variation in pressure at different positions in space at one instant of time. The resultant wave at the bottom is simply the sum of the pressure variations caused by the two top waves if present at the same time. You can produce the same pattern with the simulation by adding two waves each of amplitude +30 with the first at frequency f_1 = 2.0 kHz and the second at twice the frequency, or f_2 = 4.0 kHz. In the music business, the first wave is called the fundamental, and the second is the second harmonic, or the first overtone. The resultant complex wave produced by adding the two is the pattern seen in the simulation and reproduced below. Show that the bottom wave is just the sum of the two waves at each position of space.

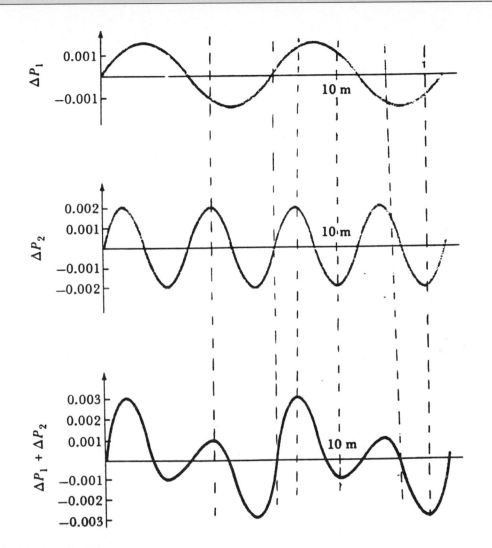

SUMMARY

- Beats are produced by adding two waves of slightly different frequencies.
- The beat frequency is the absolute value of the difference in frequency of the two waves: $f_{beat} = f_2 - f_1$.
- Complex waves are formed by adding a fundamental-frequency wave with higher-frequency harmonics that are integral multiples of the frequency of the fundamental wave.

A source (the bell in the simulation) emits a sound at the source frequency f_S. A listener or observer (the person in the simulation) detects the sound at what may be a different frequency f_L. Under what conditions are these frequencies the same? Under what conditions are they different? Which frequency is greater? Is there any useful application for this frequency difference? These questions are the subject of this Doppler-effect activity.

Note: The signs for the velocities in the simulation indicate the velocity of the source or listener relative to an axis that points toward the right.

Questions 1 - 2 — Doppler Frequency, Qualitative: For each situation described below, decide if the listener (observer) frequency is greater than, equal to, or less than the source frequency.

(a) The listener moves toward the stationary source:	$f_S > f_L$	$f_S = f_L$	$f_S < f_L$
(b) The source moves toward the stationary listener:	$f_S > f_L$	$f_S = f_L$	$f_S < f_L$
(c) The listener moves away from the stationary source:	$f_S > f_L$	$f_S = f_L$	$f_S < f_L$
(d) The source moves away from the stationary listener:	$f_S > f_L$	$f_S = f_L$	$f_S < f_L$
(e) The listener and source move in the same direction at the same speed:	$f_S > f_L$	$f_S = f_L$	$f_S < f_L$

Invent a qualitative rule (without an equation) that indicates under what conditions the listener frequency is greater than the source frequency, less than the source frequency, and equal to the source frequency. Write the rule in the space below.

Question 3: Have you ever experienced the Doppler effect? Describe the experience(s).

Questions 4 - 5 — What causes the moving-source Doppler effect? In the simulation, set $v_{listener}$ = 0.0 m/s, v_{source} = −10.0 m/s, and the source frequency f_S = 10,000 Hz. Run the simulation. You should see a pattern such as shown below but with many more wave crests. Look at the pattern below and pretend that you are looking down from above on a very large swimming pool. Each crest is a wave that was created by a large beach ball that is bobbing up and down in the water as it moves from the right to the left. The larger radius crest was created when the ball was at position 1, and the smaller radius crest when it was at position 2. The wavelength of waves leaving the source is the distance between crests. Explain in the space below why the listener frequency is greater than the source frequency when the source moves toward the listener and less when the source moves away. Think about the wavelength, and how wavelength is related to wave speed and frequency.

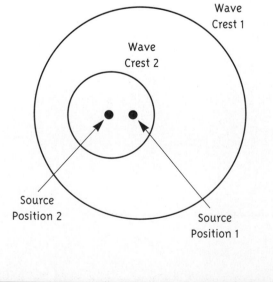

Question 6 — What causes the moving-listener Doppler effect? In the simulation, set vlistener = +4.0 m/s, vsource = 0.0 m/s, and the source frequency ƒs = 10,000 Hz. Run the simulation. Note in particular the rate at which the listener moves past wave crests. Now, change the listener velocity to vlistener = -4.0 m/s. Run the simulation again and note again the rate at which the listener moves past wave crests.

(a) If the listener was moving left at the same speed as the waves, what frequency would the listener observe? Explain.

(b) Explain in the space below why the listener frequency is greater than the source frequency when the listener moves toward the source and why the listener frequency is less than the source frequency when the listener moves away from the source.

SUMMARY

- $f_{listener} < f_{source}$ when the listener and the source are moving farther apart.
- $f_{listener} > f_{source}$ when the listener and the source are moving closer to each other.
- $f_{listener} = f_{source}$ when the listener-source separation is constant.

DOPPLER EQUATION

- $f_{listener} = f_{source} \left(\dfrac{v \pm v_{listener}}{v \pm v_{source}} \right)$ where

  v = the speed of sound in the medium (about 340 m/s for sound in air);

  $v_{listener}$ = listener speed (use + in equation if moving toward source and − if moving away); and

  v_{source} = source speed (use − in equation if moving toward listener and + if moving away).

SUMMARY

- $f_{listener} < f_{source}$ when the listener and the source are moving farther apart.
- $f_{listener} > f_{source}$ when the listener and the source are moving closer to each other.
- $f_{listener} = f_{source}$ when the listener-source separation is constant.

DOPPLER EQUATION

- $f_{listener} = f_{source} \left(\dfrac{v \pm v_{listener}}{v \pm v_{source}} \right)$ where

  v = the speed of sound in the medium (about 340 m/s for sound in air);

  $v_{listener}$ = listener speed (use + in equation if moving toward source and − if moving away); and

  v_{source} = source speed (use − in equation if moving toward listener and + if moving away).

For each of the following situations, predict the listener (observer) frequency. After your prediction, adjust the sliders to see how you did.

Question Number:	1	2	3	4
Listener velocity (m/s)	0	0	+4.0 (right)	−4.0 (left)
Source velocity (m/s)	−10 (left)	+10 (right)	0	0
Source frequency (Hz)	10,000	10,000	10,000	10,000
Predicted frequency (Hz)				

Question 7: Two students decide to test the Doppler effect by walking along a street as they blow identical 1000-Hz whistles. Student One is ahead and moves right at speed 1.0 m/s. Student Two (initially on the left) moves right at 4.0 m/s. (a) What beat frequency does Student Two hear when playing her whistle and while still behind Student One? (b) What beat frequency does Student Two hear after passing Student One? Make your predictions. Then check your predictions by running the simulation.

Why does a violin playing concert A at 440 Hz sound different from a flute or a clarinet playing the same frequency sound equally loud? The answer is in part due to what we call the harmonic content of the complex waves produced by these musical instruments. A complex wave consists of one wave at a fundamental frequency — such as the 440 Hz frequency of concert A played by a violin. The wave also consists of other waves at integral multiples of the fundamental. All of these waves are added together to form the complex wave. In this activity, you can construct some very special complex waves by adding the fundamental and the right combination of higher harmonics.

Construct a Sawtooth Wave:

• Open the simulation, which displays the wave pattern that looks like the teeth of a saw (upper right box).

• Leave INITIAL SIGNAL checked.

• Click on SYNTHESIZED SIGNAL to remove the check mark.

• Move the slider for n the floating window to 1. (You may have to scroll down the simulation window to see this slider.)

• Now check SYNTHESIZED SIGNAL. You will see a sine wave at the same frequency as the sawtooth wave.

Finish the Sawtooth Wave:

• Change the slider to n = 2. You see a combination of the fundamental (n = 1) and the second harmonic (n = 2) superimposed on the sawtooth wave.

• Click the slider again to add the n = 3 wave to the other two waves.

• Keep adding more harmonics to produce a wave that looks very much like a sawtooth wave.

A sawtooth wave can be produced by combining a fundamental-frequency wave with a carefully selected combination of higher frequency harmonics.

Other Complex Waves: You can repeat this activity with other types of wave patterns. They are formed by different amplitude combinations of the fundamental and its higher harmonics.